ANALYSIS OF REACTOR VESSEL RADIATION EFFECTS SURVEILLANCE PROGRAMS

Prepared under the auspices of
THE METAL PROPERTIES COUNCIL
by L. E. Steele and C. Z. Serpan, Jr.

ASTM SPECIAL TECHNICAL PUBLICATION 481

List price $26.00

AMERICAN SOCIETY FOR TESTING AND MATERIALS
1916 Race Street, Philadelphia, Pa. 19103

Library of Congress Catalog Card Number: 70–129930
ISBN 0–8031–0066–3

NOTE

The Society is not responsible, as a body,
for the statements and opinions
advanced in this publication.

Printed in Baltimore, Md.
December 1970

Foreword

Many factors have combined to accelerate the programs of the members of the electric power industry for the erection of power stations using nuclear fuel.

The Metal Properties Council was assigned the role of analyzing and up-dating the mass of data expected to be accumulated as a result of the surveillance testing program that has been adopted as a means of measuring the changes in properties of pressure vessel materials as a function of neutron fluence.

In preparation for this task, L. E. Steele and C. Z. Serpan, Jr., have written this volume which summarizes and analyzes all pertinent information concerning the influence of neutrons on the properties of pressure vessel materials. As a convenience to the electric power industry, the Metal Properties Council previously has circulated a brief report on "Suggested Guidelines for Selecting a Surveillance Specimen Evaluation Laboratory."

This volume has been reviewed and approved by Subcommittee 6 of the Metal Properties Council. It is published with the cooperation of the American Society for Testing and Materials.

An attempt will be made to continue to gather information resulting from pressure vessel surveillance test programs in power reactors. The results of the analysis of this program also will be prepared by the Metal Properties Council.

Adolph O. Schaefer

The Metal Properties Council

Related ASTM Publications

Effects of High-Energy Radiation on Inorganic Substances, STP 400 (1966), $5.25

Irradiation Effects in Structural Alloys for Thermal and Fast Reactors, STP 457 (1969), $36.00

Contents

Synopsis and Report Highlights 1

CHAPTER 1—Basis for Vessel Surveillance (Figs. 1–15) 17

CHAPTER 2—Neutron Embrittlement of Pressure Vessel Steels—A Brief Review (Figs. 16–46) (Tables 1–14) 47

CHAPTER 3—Army Reactor Vessel Surveillance and Vessel Examination (Figs. 47–63) (Tables 15–21) 103

CHAPTER 4—Commercial Reactor Vessel Surveillance (Figs. 63–97) (Tables 22–39) 137

CHAPTER 5—Neutron Dosimetry and Spectrum (Figs. 98–102) (Tables 41–44b) 207

CHAPTER 6—Surveillance Programs Critique and Recommendations (Figs. 103–105) (Appendix I, Surveillance Program Recommendations) (Appendix II, Suggested Guidelines for Selecting a Surveillance Specimen Evaluation Laboratory) 237

Glossary 261

Index 267

SYNOPSIS AND REPORT HIGHLIGHTS

This summary section is provided to aid the reader in finding specific aspects of reactor pressure vessel surveillance within this report and to provide a brief review of its major highlights. For clarity, the synopsis is divided to correspond with the six chapters of the report.

1. Basis for Vessel Surveillance.
2. Neutron Embrittlement of Reactor Vessel Steels—A Brief Review.
3. Army Reactor Vessel Surveillance.
4. Commercial Reactor Vessel Surveillance.
5. Neutron Dosimetry and Spectrum.
6. Surveillance Programs Critique and Recommendations.

In addition to this overall review of report highlights, the introduction to each chapter provides an abbreviated review of the contents of that chapter.

CHAPTER 1—Basis for Vessel Surveillance

Introduction

This report was undertaken at the request of the Metal Properties Council to review and analyze available reactor pressure vessel radiation damage surveillance data in order to guide new reactor owners and others having an interest in this aspect of nuclear power. In Chapter 1, some important factors are outlined relative to the basis for reactor vessel surveillance.

Highlights

1. Ferritic steels such as those used for nuclear power reactor pressure vessels undergo a marked transition from ductile-to-brittle fracture be-

havior over a relatively narrow temperature range (100 to 150 F) usually near 0 F.

2. Neutron radiation gradually elevates the transition temperature range, lowers the energy required to fracture, hardens, and increases the strength of a steel.

3. These effects of radiation tend to exaggerate the chances for brittle fracture of the vessel; hence, the need for some technique to measure, progressively, radiation effects and the propensity of the vessel steel to serve its function with assured reliability.

4. The usual basis for assessment of the vessel's condition in service is to expose notched impact bars (Charpy V-notch) of the vessel steel (plate and weld) at the vessel wall during operation and to remove and test groups of these specimens periodically. Radiation induced changes in the specimens are accepted as a measure of change in the vessel and thus provides a basis for reactor operational decisions which help assure vessel reliability in service.

5. The changes considered most significant are the increase in transition temperature and the reduction in fracture energy, especially the energy absorbed at temperatures where ferritic steels exhibit full shear fracture, usually associated with high ductility, but, with severe irradiation, may, in fact, permit low energy tearing. In the latter situation, assuring higher vessel temperatures before significant pressurization may be of no consequence. These two types of changes often are referred to as radiation embrittlement. The term embrittlement is relative and is used here to denote either small or large changes. It is important in either context because of the important function of the vessel and because of the economic costs of overly conservative design or of operational limitations which result from uncertainty.

6. The Charpy V-notch specimen forms the primary basis for vessel surveillance and for measuring radiation effects on notch ductility. This test is made more meaningful by the fact that, when properly correlated with the drop weight test, a nil-ductility transition (NDT) temperature can be determined for a given vessel steel. The use of NDT is important because it gives a *point* for measuring an increase in transition temperature rather than a *range*, and, since it was originally developed on the basis of structural failures, it provides a reference point for fracture analysis using the fracture analysis diagram (FAD) for relating stress, temperature, and flaw size for failure. An NDT or Charpy V-notch based increase in transition temperature is valuable but does not describe the radiation induced reduction in energy absorption at the ductile (shelf) end of the transition range. (Criteria are now being developed which will guide vessel embrittlement projections using the Charpy V-notch data (correlated with fracture toughness data) so as to integrate both reduced shelf energy absorption and transition or NDT temperature increases.)

7. Recent developments relative to reactor pressure vessel safety include industry programs for describing statistically the properties of representative vessel steels and for inspecting before and during service for critical flaws. A related United States Atomic Energy Commission (USAEC) sponsored program seeks to evaluate the potential for failure resistance in heavy section steels.

8. Within the USAEC program, a most significant development has been recorded on the basis of full thickness plane strain fracture mechanics K_{Ic} tests and dynamic tear (DT) tests. Both tests demonstrate a dramatic upsweep in fracture energy representing the ductile-brittle transition range, thereby demonstrating that rising ductility with increasing temperature overcomes the tendency of the high mechanical constraint in heavy section to suppress significantly the inherent ductility of the steel as had been suggested by certain theoretical studies.

9. Static linear elastic fracture mechanics tests which seek to define a critical stress intensity value at an artificial sharp flaw in a specimen of given dimensions were applied using progressively larger specimens (1 to 12 in.). The result was a curve sweeping rapidly up at about the NDT temperature and lying about 60 to 70 F below the rapid upsweep defined by the dynamic tear test results conducted using 3, 6, and 12-in. specimens. Thus, the upper temperature for valid K_{Ic} tests (using full thickness tests) lies no higher than the temperature of NDT +40 to 60 F which normally corresponds to the upper-middle region of the transition as defined by the Charpy V-notch curve.

10. Several important observations are noted from the dynamic-tear test results:

(*a*) With increasing specimen size a progressively sharper transition curve is defined, reflecting increased mechanical constraint in the larger specimens.

(*b*) DT specimens of 3, 6, and 12 in. thickness all define similar transition curves which rise sharply only after a temperature of approximately 120 F above the NDT temperature. Thus, a thickness effect (elevation of the transition temperature of about 70 F) is noted. This is most important since it demonstrates the need for accepting a criterion based upon the Charpy specimens tests which will assure vessel ductility in spite of this thickness induced elevation of the transition range. Criteria are now being considered by the USAEC to provide such assurance under any projectable vessel condition based upon surveillance data. New criteria will depend upon the ductility inherent in steels at a point above about half the energy at the midtransition point or a related dynamic fracture toughness value for describing full shear (shelf level) fracture resistance.

(*c*) The DT test results generally portray a transition which is higher than that provided by the Charpy V-notch test results, thereby suggesting that a somewhat more conservative basis is provided by the former. However, the new criteria mentioned in (*b*) will give consideration to a more conservative and quantitative interpretation of Charpy V-notch data for defining notch ductility of vessel steels.

(*d*) Gross plastic loading is required to fracture DT specimens at temperatures representing the ductile shelf of the transition curve. An energy of 760,000 ft·lb (full machine capacity) was absorbed by an A533–B steel specimen having an unfractured cross section of about 12 by 12 in. It is estimated that, at the test temperature (215 F), energy of about 1,000,000 ft·lb would have been required for full fracture.

(*e*) These results suggest that reasonable assurance against vessel failure can be obtained by following the dynamic tear results and a related conservative application of Charpy V-notch surveillance data. Similarly, projection of critical flaw sizes for failure at the temperatures involved in reactor vessel applications requires flaws measured in feet rather than inches.

(*f*) With severe irradiation (not expected under normal reactor conditions), however, the possibility of low energy tearing overshadows these transition temperature aspects. In this regard, properties of welds and of plates in the weak orientation (transverse to primary rolling direction) are the most critical. Data for this orientation must be developed to provide knowledge which will permit action which obviates potential consequences of low energy tearing.

CHAPTER 2—Neutron Embrittlement of Pressure Vessel Steels—A Brief Review

Introduction

This chapter was written as background for specific reactor vessel surveillance results to help explain, using experimental results, why radiation damage surveillance is needed.

Studies have been and are being conducted to assess the effects of high energy nuclear radiation upon the properties of reactor vessel steels under conditions which simulate nuclear service. Selected highlights of these program results are outlined.

Highlights

1. Neutron embrittlement of vessel steels varies from very small to very large depending upon a number of factors which can be classified as

environmental or materials related. The most significant environmental factors are the exposure temperature and neutron fluence, and the most significant materials factors are the composition and microstructure of the specific vessel steel of interest.

2. Neutron fluence or exposure level for steel specimens exposed at a given temperature determines how much embrittlement will be sustained. Embrittlement is highly nonlinear with very rapid changes to about 1 to 2×10^{19} neutron (n)/cm^2 > 1 MeV with progressive deceleration of change to about 7 to 8×10^{19} where there are indications of damage saturation. With elevated temperature irradiation ($\geq$550 F) this pattern occurs at lower fluences.

3. Studies conducted on one heat of A302–B steel in different types of reactors have shown that the distribution of neutrons by energy (neutron spectrum) may affect significantly the embrittlement results giving more embrittlement in some cases and less in others than would be expected by assuming a spectrum and conducting the usual activation dosimetry. Further, in certain environments, significant errors in judging expected embrittlement have resulted from the conventional use of fluences in terms of neutrons >1 MeV. Thus, it is suggested that the vessel wall spectrum be computed and considered to lower energy levels (for example, 0.5 MeV) for defining damage per unit flux at the reactor vessel. (Advancements in neutron dosimetry and spectral analysis, which are reviewed in Chapter 5, can be expected to continue with the ultimate result that new standards for defining the damaging neutron environment will result).

4. Irradiation at higher temperatures results in significantly reduced neutron induced change as a result of thermal annealing processes which become quite evident and progressively more advantageous as the irradiation temperature is raised above about 450 F.

5. Advantage may be taken of the potential for "annealing-out" neutron damage through the elevation of the temperature above the normal reactor vessel operating or exposure temperature. The degree of thermal correction or annealing achieved depends largely upon the same factors which influence embrittlement plus the conditions of annealing; that is, annealing is most effective if carried out much above the steel exposure temperature for an extended period. Further, annealing effectiveness appears to be related directly to the ease with which the steel is embrittled. For example, in steels showing a high propensity for neutron embrittlement, annealing is highly effective and vice versa. Similarly, steels irradiated at low temperatures respond more favorably to annealing than steels irradiated at elevated temperature even if the same temperature increment (say 150 F) above the exposure temperature is applied for correction in both cases.

6. Specimens of pressure vessel steels have been irradiated in test reactors under applied loads to simulate reactor service. Results of these

relatively short term treatments have indicated no significant effect of applied stress upon the radiation induced change. Longer term tests of this factor are underway with fracture mechanics specimens exposed in vessel surveillance programs under static loads.

7. Very large differences in irradiation sensitivity have been observed between different steels, between heats of the same steel, and between forms or conditions (as weld to plate) of the same type of steel. The magnitude of these differences has been so great in some comparisons as to overshadow the effects due to the normal environmental variations. Variations in embrittlement sensitivity between steels have ranged as high as an order of magnitude as has the variation between weld and plate. The more usual situation, however, involves a factor of 2 to 3 difference in sensitivity from best to worst.

8. The most significant factor in radiation embrittlement sensitivity is composition though microstructure has been shown to affect the level of change as well. The observation of low and high sensitivity to radiation embrittlement in commercial steels has led to studies of special laboratory heats in which copper, phosphorus, and vanadium have been identified as critical to the level of sustained embrittlement.

9. The identification of critical elements in radiation embrittlement has been verified empirically by studies of composition and embrittlement in many steels and weldments and recently through special production plates and welds in which specifications were applied to minimize the undesirable elements. Control of copper (to less than about 0.080 percent) and phosphorus (to less than about 0.010 percent), and general minimization of other residual elements (such as arsenic, lead, antimony, tin, sulfur, etc.) provides assurance of low radiation embrittlement sensitivity of reactor pressure vessel steels.

10. Studies of radiation embrittlement sensitivity, while incomplete, have progressed to the point where it appears that carefully applied composition specifications can assure routine production of radiation insensitive pressure vessels. Further, these studies have provided important guidelines for projecting in advance the level of radiation embrittlement which may be expected in a particular reactor vessel.

11. Radiation effects significant changes in tensile properties, increasing the yield and tensile strength and reducing the elongation (total and uniform), the strain hardening exponent, the TS/YS ratio, and the reduction of area. At the usual reactor operating temperatures (>500 F), however, these changes do not appear to be significantly detrimental at neutron fluences below about 3 to 5×10^{19}.

12. The fatigue strength of vessel steels do not appear to be affected significantly by radiation. However, the important parameter of fatigue crack propagation rates in nuclear versus nonnuclear environments is yet to be studied.

These significant experimental aspects of radiation effects on pressure vessel steels provide a strong basis for meaningful comparison and correlation of data from individual reactor vessel surveillance programs.

CHAPTER 3—Army Reactor Vessel Surveillance

Introduction

This chapter was written primarily because of the advanced exposure condition in the compact Army reactors and the accelerated embrittlement experience which some day may be directly applicable to the larger commercial nuclear power plants. Of special interest is the in-place annealing of the SM–1A reactor vessel and the postservice testing of the PM–2A reactor vessel. An important consideration is the fact that these two Army reactors were made up of forged rings, thereby eliminating welds in the highest flux regions of the vessel.

Highlights

1. The compact Army reactors provide early accumulation of very significant, if not severe, nuclear exposure levels; thus, they offer advance knowledge of conditions which may eventually accrue to the larger reactor vessels.

2. The SM–1A reactor vessel surveillance was limited by space to the placement of dosimeters along the wall and specimens above the core. Supplementary experiments using test reactors, however, provided the needed background on the radiation embrittlement sensitivity of this vessel steel and, with the dosimetry, provided the basis for later application of the first in-place annealing (to correct radiation damage) of any reactor vessel. The SM–1A vessel steel, ASTM Type A350–LF1 (modified with added nickel), proved to be very sensitive to radiation at the temperature of reactor vessel operation (~430 F) but also proved very responsive to thermal correction by annealing at a temperature of about 140 F above the operating level.

3. The SM–1A was annealed to correct embrittlement after the vessel reached a neutron fluence of 1.6×10^{19} n/cm^2 (assumed fission spectrum) in 40.4-MW years of operation and caused an increase in transition temperature (ΔT) of about 260 F. This ΔT was 80 F short of the Army imposed limit of 340 F. The Army's conservative basis for projecting the annealing point along with superimposed AEC conservatism resulted in

annealing before limits based upon experimental data would have required. Annealing recovery was equal to about 70 percent of radiation induced ΔT or ΔNDT after annealing for one week at a temperature of 572 F. This low annealing temperature was made possible by the very favorable response to heat treatment at about 140 F above the operating temperature. This permitted annealing using nuclear heat (low operating power with throttled coolant flow) which allowed the needed rise in temperature with a relatively small related rise in operating pressure.

4. It was required that neutron spectrum and fluence calculations be performed and projected on the basis of $n/cm^2 > 0.5$ MeV. This exercise indicated the use of $n/cm^2 > 0.5$ MeV to be the more favorable treatment for the SM–1A reactor because the relative abundance of neutrons having energies greater than 0.5 MeV was lower in the SM–1A spectrum than in the Watt fission spectrum.

5. In spite of the success of the SM–1A vessel annealing sequence as verified by experiment and by specimens simulating the irradiation and annealing which were removed from the SM–1A after annealing, the high vessel wall flux requires planning for subsequent repeat annealing cycles. Experiments have been conducted which validate this plan. The absence of surveillance specimens at the vessel wall complicates planning and dictates continued conservatism toward vessel embrittlement however.

6. A similar sister reactor to the SM–1A, the SM–1, has not accumulated the high vessel exposures of the SM–1A because of a smaller core (yielding a lower vessel wall flux) and reduced time at power because of its training function. Nevertheless, concern for the embrittlement to a point requiring annealing has resulted in analytical evaluations of the vessel exposure flux, fluence, and spectrum in an effort to project the vessel condition in future years.

7. The PM–2A reactor vessel was tested nondestructively and destructively (as a single specimen and as a source of materials for multiple laboratory specimens) after the reactor had served its function at Camp Century, Greenland, in the early 1960's.

8. Nondestructive tests of the PM–2A included ultrasonic examination which showed no significant included flaws in the vessel, out-of-roundness tests which showed none which would significantly accentuate stresses in planned destructive tests, and gamma ray spectrometry which indicated (unfortunately erroneously on the high side) the neutron fluences sustained at various positions around the vessel.

9. Destructive testing of the PM–2A vessel was to show how fracture mechanics would predict the flaw, stress, and temperature conditions for failure. Predictions proved to be in error, but the test sequences were complicated by the errors in projected fluence, the gradient in exposure through the vessel wall, the significant portion of ductile Type 304 stainless steel lining in the remaining vessel ligament at failure, the complex cyclic

test procedures which involved loading and unloading at various temperatures, fatigue cycling to sharpen the machined flaws, and, finally, stress corrosion to sharpen the machined flaw. In spite of all these difficulties, the full vessel test suggested a very reliable component in that failure required higher stresses than had been predicted.

10. Laboratory tests of the failed PM–2A vessel showed:

(*a*) Increases in transition temperature in line with experimental data projections for the steel and its service condition.

(*b*) Tensile ductility not greatly affected by the service environment ($\sim 1 \times 10^{19}$ n/cm^2 > 1 MeV at 510 F).

(*c*) Neutron fluences across the vessel wall thickness presenting a linear relationship on a semilog plot which should aid similar projections for thicker vessels.

(*d*) Fracture mechanics data providing relatively good agreement with the failure conditions though the possibilities for projection to temperatures having meaning in an operational sense were shown to be quite limited.

11. In general, the small Army reactors lead their larger commercial counterparts in the problems which may be anticipated due to neutron radiation damage of the vessel.

CHAPTER 4—Commercial Reactor Vessel Surveillance

Introduction

This is a key chapter in that it meets the primary objectives of this report to describe and analyze results of commercial reactor vessel surveillance programs. Since Chapter 6 provides a critique of each individual reactor program, emphasis in this synopsis is upon description, interim results, and relative status of programs.

Highlights

1. Only five commercial reactors have operated long enough with surveillance programs to have produced significant results. These include the General Electric constructed reactors: Big Rock Point, Dresden 1, and Humboldt Bay; the Westinghouse constructed reactor–Yankee–Rowe; and the Allis–Chalmers constructed reactor–Elk River. In spite of the limited number involved, these reactors as pioneers in surveillance (and in many other aspects) provide valuable data for viewing the future; hence, a detailed analysis is of special value.

2. Information describing individual surveillance programs of 25 com-

mercial reactors are tabulated along with vessel material, planned life, anticipated maximum vessel fluence, assumed plant factor, and projected maximum vessel NDT.

3. A comparison of the range of anticipated vessel fluences shows a high of 6.0×10^{19} n/cm^2 > 1 MeV for Shippingport and a low of 1×10^{17} for Vermont–Yankee. A similar extreme of projected maximum NDT values ranges from 438 to 40 F for these two reactor vessels.

4. Projected fluence values of some of the earlier reactors have been in error on the low side, however. For example, calculated values for Big Rock Point were about 3×10^{18} > 1 MeV while measured values (discussed in detail in Chapter 6) are 3×10^{19} > MeV. Similar, though smaller, variations have been determined for other plants including Dresden 1 and Yankee–Rowe. Related errors in maximum NDT temperature are predicted.

5. Surveillance program plans reported to date are largely those of General Electric and Westinghouse, which follow the ASTM Recommended Practice, E 185 (Recommended Practice for Surveillance Tests on Structural Materials in Nuclear Reactors), in general, but are expanded over this in several specific instances. Standard General Electric and Westinghouse programs are described. Variations from the standard are prevalent in General Electric's case as those vessels having electroslag welds receive special attention and those which have anticipated fluences $<5 \times 10^{17}$ receive minimum treatments unless an electroslag weld is present in the belt region of the vessel. Standardization is developing rapidly, but many early plants contained no weld or weld heat affected zone specimens (a potentially serious omission).

6. Surveillance programs contain predominantly Charpy V-notch and tension specimens (in widely varying total numbers), but some contain fatigue specimens (Elk River) and some fracture mechanics specimens (certain Westinghouse and Babcock and Wilcox produced reactors). While the type of specimen used is most important, the inclusion of specimens truly representing the vessel materials is even more important.

7. Three surveillance capsule positions normally are used: just inside the vessel wall, an accelerated (usually inside thermal shield) exposure position, and a thermal control (out of significant flux) position. The latter two are not required but provide guidance to aid the projection of vessel wall surveillance results to later years with confidence. Capsules usually contain, in addition to vessel steel specimens, reference (correlation monitor) specimens of a well known steel heat, low melting point or eutectic alloys to provide guidance as to temperatures attained plus neutron dosimeters to aid in determining the flux and fluence attained in the capsule exposure period. Dosimeters are also valuable for projecting NDT temperatures in future years provided the assumption of no significant change in core loading or power density is valid.

8. The Big Rock Point reactor surveillance program was well planned; however, the determination of the actual vessel flux was an order of magnitude greater than calculated and left the program inadequate for the full life of the plant. It also lacked specimens of a well-known reference steel. During 1969 the last surveillance specimens were removed and evaluated. Fortunately, the availability of specimens in accelerated locations, and the presence of plate, weld, and weld heat affected zone (HAZ) specimens which could be removed at several year intervals provided the basis for reasonably good lifetime projections. This program shows the importance of having extra surveillance specimens available plus some means for adding them. The unexpectedly high flux made it necessary that specimens be removed much ahead of the planned removal, thereby depleting in only 6 years what was intended to cover approximately 30.

9. The Yankee–Rowe vessel surveillance program was rather limited containing ten Charpy V-notch and tension specimen capsules, eight of which were in accelerated and two in vessel wall irradiation locations. The program was further limited by the loss of four accelerated capsules and one vessel wall capsule due to failure of the capsule anchor assembly. All four accelerated capsules were removed after only one core life (Core II) and the single wall capsule was removed after three cores (Cores II, III, and IV). The specimens were prepared only from vessel plate (no weldment specimens), and the surveillance plate was separately heat treated so its properties did not meet exactly those of the vessel. In spite of these several limitations, careful dosimetry analysis and full evaluation of the well-known reference steel provided the key to meaningful projection of later life conditions in the Yankee vessel. Many lessons are inherent in the results of the Yankee–Rowe surveillance.

10. The Dresden 1 program was very large and though a similar underestimation (factor of four) of flux and fluence occurred, the availability of many specimens and a means for adding more if needed proved very far-sighted. Thus, this second oldest U.S. nuclear power plant has an on-going surveillance program which, with careful analysis (for specimens added after the initial startup), can and will provide a continuing and acceptable vessel radiation damage surveillance program. One weakness was the lack of a well-known reference steel. Fluences projected from the early measured dosimetry results made it important to continue surveillance even if on a delayed insertion basis. The reactor owner has been especially diligent in seeking full answers to the radiation damage question.

11. The Elk River reactor vessel surveillance program is large (40 capsules) but suffers the rather severe handicap of not containing specimens of the actual vessel material. Instead, it includes three heats of A302–B steel (neither truly representative), two of A105, and one of A212–B. In addition, certain transition welds, A105 to Type 304 stainless, were exposed in Elk River. Another early limitation of this program was the lack of

definitive dosimetry; this has been corrected in recent efforts. Also, difficulty in exact measurement of temperature has hampered analysis of results in the Elk River case.

A unique aspect of the Elk River surveillance was the inclusion of fatigue specimens. This was instigated by the observation of small cracks in the stainless steel vessel overlay. Fatigue results have been obtained.

12. The Humboldt Bay vessel surveillance is in a relatively infant state with accelerated irradiation position specimens having been tested after two years of operation. Many capsules containing specimens from plate, weld, HAZ, and from the Dresden steel provide an ample basis for good surveillance; results to date are too limited for really significant conclusions.

CHAPTER 5—Neutron Dosimetry and Spectrum

Introduction

This chapter provides a description of the neutron dosimetry and spectrum analysis techniques necessary to meaningful reactor vessel surveillance. Without good neutron dosimetry and spectrum analysis misleading conclusions may result even with the best materials surveillance. Basic definitions and accepted techniques are described.

Highlights

1. As early surveillance data are used to project properties changes in the vessel in later life, success depends upon knowing the number or density and the energies of neutrons which pass through a given cross section of the steel vessel; this is neutron dosimetry and spectrum analysis.

2. Neutron dosimetry is accomplished through the placement of wires or foils of selected metals or alloys for which the action of neutrons in causing activation and secondary radiation emissions is well known. Such materials, known as dosimeters, monitors, or detectors, react to neutron bombardment by emitting radiation of a particular type; the intensity coupled with the exposure period provides an indirect measure of neutrons which have passed through the detector.

3. Since single detectors are responsive to neutrons of a particular energy range, proper dosimetry depends upon knowing the response (activation) of a group of detectors so as to assure coverage of the spectrum at the surveillance location. Thus, the best approach involves use of several well-known detectors which are activated by neutrons of widely different

energy levels. No such group of detectors remotely approaching an ideal response has yet been found for long-term applications such as reactor vessel surveillance. A listing of the preferred detectors and their pertinent nuclear characteristics is provided, however. In addition, it is necessary that a computation of the spectrum at the vessel wall be used as a basis for selecting the proper activation constants.

4. Computation of the spectrum at the location of surveillance specimens is essential as it provides the basis for use of the proper dosimeter activation cross-section value, permits a reasonable evaluation of the contribution to damage by various components of the spectrum (thermal versus intermediate versus high energy, etc.), and provides a good test of the reported fluence in relation to observed embrittlement.

5. Recent developments in computer technology provide sophisticated techniques for computing the neutron energy spectrum at any reactor location and even alternate approaches for check evaluations. Developments in neutron dosimetry have provided a series of detectors which cover the general range of energies of interest for steel irradiation damage studies and thereby permit a dosimetry validated spectrum computation. Use of a damage function (cross section for damage production) for a given spectrum further aids the analysis. Damage function development depends upon a computed spectrum augmented by dosimetry and, where possible, by changes in properties in a particular steel. Computer analysis of such data represents the latest and best approach to radiation damage analysis.

6. A more complex spectrum analysis was applied and is reviewed for the SM–1A reactor surveillance. Due to lack of space in the SM–1A reactor, it was necessary to expose representative materials in a test reactor, the low intensity test reactor, to simulate exposure of the SM–1A vessel and then to calculate and adjust for spectrum differences. This exercise provided valuable experience and added confidence in the use of computed spectra in conjunction with measured changes in properties and good neutron dosimetry at the reactor positions of interest.

CHAPTER 6—Surveillance Programs Critique and Recommendations

Introduction

This chapter outlines conclusions as well as critiques of results from various reactors which have thus far produced surveillance data. Upon this background, important features or factors to be considered in producing a comprehensive vessel surveillance program are outlined. Since the Army reactor programs have been reviewed in the synopsis of Chapter 3, no further highlight discussions are provided. A brief critique is provided for each of the five commercial reactors, however.

Highlights

1. Some overriding general comments about what a good surveillance program should contain include the following:

(*a*) It should provide for the irradiation of steel (plate, weld, and HAZ) from the actual pressure vessel.

(*b*) It should utilize an irradiation location and condition which duplicates as closely as possible the physical and nuclear conditions of the vessel.

2. Other important factors include:

(*a*) The assurance that specimens are completed in accordance with pertinent standards.

(*b*) That these specimens are tested on carefully calibrated test apparatus.

(*c*) That the nuclear constants used in neutron dosimetry and spectral analysis are the best and latest available.

(*d*) That a complete record of the materials history is recorded and retained (along with archive segments of the actual vessel steel) for future analysis.

3. Critique of Big Rock Point Reactor Vessel Surveillance

(*a*) Materials were of the actual vessel and weldments though certain heat treatments were simulation of the vessel.

(*b*) The number of vessel wall specimens, thought to be adequate for the projected lifetime fluence of 3×10^{18} n/cm^2 > 1 MeV, proved inadequate for good lifetime coverage in view of the order of magnitude higher measured fluence (3×10^{19} projected for vessel lifetime).

(*c*) The mode of specimen notching in some cases was nonstandard (parallel rather than perpendicular to the surface) and missed the fusion zone in some HAZ specimens.

(*d*) Nevertheless, the program was complete enough to permit projection of a lifetime fluence and NDT (~200 F) in the weld metal (the worst case), which for the rather high (~575 F) operating temperature should not present more than an operational control question in later years of reactor life unless core changes produce higher fluences or for some reason the operating temperature is reduced significantly.

4. Critique of Yankee–Rowe Reactor Vessel Surveillance

(*a*) The strongest feature of this effort was the inclusion of a well known A302–B reference steel; the weakest factor was the mode of capsule attachment which resulted in the loss of half of the ten capsules and a questionable positioning of the remainder.

(*b*) This program lacked weld specimens and the plate specimens were from a separately heat treated plate.

(*c*) The lack of neutron dosimeters and temperature monitors, especially the latter, also weakened the program.

(*d*) In spite of the noted weaknesses, however, the availability of the well characterized reference steel saved the effort, making possible reasonably good vessel embrittlement projections. These projections (at most an NDT of 305 F), assuming exposure conditions to continue unchanged, or if changed, changed to minimize radiation damage, suggest that nothing more than a conservative operating practice need be applied. Doubt about the weld sensitivity and the operating temperature, nevertheless, requires conservative interpretation and operational practice.

(*e*) It should be noted in favor of this effort that all facets were well documented and good records were maintained, though the availability of surplus "archive" specimens or material was very limited.

3. Critique of Dresden 1 Reactor Vessel Surveillance

(*a*) This pioneer program was initially lacking in weldment specimens and in temperature monitors, but this was corrected and the Dresden became the most carefully monitored of the early power reactors.

(*b*) One aspect of specimen preparation, however, creates some doubt for the interpretation of data. Charpy specimens were taken from top and bottom layers of the vessel steel. These areas of quenched and tempered steels usually are superior in strength and ductility to the major section of the vessel. Further, specimens are notched contrary to standard practice, that is, parallel rather than perpendicular to the plate surface direction. Nevertheless, any significant departure from expected results would be determined by the vast program. (Dresden has been used somewhat as a test reactor for evaluating a number of vessel steels in addition to that of the Dresden vessel).

(*c*) Calculated projections of neutron flux levels were found to be low by a factor of four when measurements were made. Even the higher measured fluxes are expected to produce an NDT of only about 265 F at most. Thus, while requiring care in devising suitable operating practice for avoiding approaching NDT (or NDT + 60 F) while the vessel is significantly stressed, nothing more than a careful reactor operating plan appears necessary provided there are no changes to increase fluence or decrease temperature.

4. Critique of Elk River Reactor Vessel Surveillance

(*a*) The Elk River program violates the two major requirements of vessel surveillance: (1) the use of representative materials and (2) the placement of specimens at the vessel wall.

(*b*) The large scope of the program helps somewhat to cover these major weaknesses, however, in that a number of steels and welds along with good neutron measurement (of the wall in later phases of capsule insertion) make up the ongoing surveillance effort.

(*c*) Fatigue specimens exposed in the Elk River reactor will test this property on a prepost test basis for materials simulating plate and welds in the vessel.

(*d*) Projections of NDT at or near planned end of life (20 years) of 420 F (ΔT of 320 over initial 100 F) suggest the need for interim operational controls and continuing evaluation of the vessel condition and the possibility for correcting embrittlement by annealing if the plant is to operate anywhere near the planned lifetime.

5. Critique of Humboldt Bay Reactor Vessel Surveillance

(*a*) The materials and surveillance locations meet the primary criteria in that representative plate, weld, and HAZ are located at positions to test the effects in the actual environment.

(*b*) The specimens, however, are taken from the plate surface, thereby possibly giving misleadingly high initial properties.

(*c*) Early accelerated exposure data suggest no materials sensitive to radiation, though relatively low Charpy V-notch ductile "shelf" results for the HAZ specimens suggest a need for continuing examination of this factor.

(*d*) No dosimetry data are yet available from the vessel wall; however, comparison with the Big Rock Point reactor layout indicates no particular reason for concern for the vessel in its 40-year life. This tentative conclusion must be tempered by the results of future surveillance tests.

CHAPTER 1—BASIS FOR VESSEL SURVEILLANCE

Introduction 19

Fast Fracture of Steels 20

Background 20
Nil-Ductility Transition (NDT) Temperature or Energy Absorption Limit 23

NDT and Reactor Vessel Surveillance 25

Recent Developments in the Study of Steel Fracture 28

AEC Industry Programs 29

Industry Cooperative Program (ICP) 30
Industry Funded Program (IFP) 30
AEC-Heavy Section Steel Technology Program (HSST) 31

Advances in Fracture Evaluation Techniques 33

Fracture Mechanics 33
Dynamic Tear Test 37
Thickness Effects 42
Dynamic Tear Test Irradiations 45

References 45

CHAPTER 1—Basis for Vessel Surveillance

Introduction

The Metal Properties Council (MPC) at the request of several of its member corporations established in 1966 a Task Group to study the availability, applicability, and interpretability of data on radiation effects in reactor structural materials. The Task Group, which first met in November 1966, gave careful study to the needs in this area, especially as related to the needs of electrical utility companies newly embarked upon programs for power generation from nuclear fuel.

After several meetings, including discussions with related committees of the American Society of Mechanical Engineers (ASME) and the American Society for Testing and Materials (ASTM), it was concluded that an important task was to be found in the area of surveillance data for radiation effects on reactor pressure vessel steels. Accordingly, the MPC established a subcommittee which in turn appointed an analyst to make a study of having the following general objectives:

1. To accumulate available surveillance data and analyze these data within the framework of existing criteria for its application and with reference to experimental steel irradiation effects data.
2. To provide a critical review of existing surveillance procedures for pressure vessel surveillance with recommendations for improvement where necessary.
3. To develop and recommend a set of guidelines for choosing laboratories for testing and evaluation of control and irradiation surveillance specimens.

It was believed that such an analytical study would provide, under one cover, a summary or state-of-the-art review of the subject which would be of value to all segments of the nuclear industry. Thus, it is hoped the data from each reactor vessel surveillance program will be maximized and, at the same time, provide the basis for optimizing the return from present and future surveillance programs. Thus this study is expected to supplement individual steel irradiation effects studies, as well as the efforts of ASTM Committee E-10 on Radiation Effects and Radioisotopes in its standards activities and those of the ASME Research Committee on Effects of Radiation on Structural Materials in its efforts to stimulate the development of data useful in reactor design.

In order to meet this objective, five major topics must be analyzed and integrated to provide a meaningful analytical report. These include:

1. The current (and projected) approach to the fracture of steels.
2. Analysis of data on radiation effects in steels in terms of the critical parameters: neutron exposure (flux, fluence, spectrum), the exposure temperature, and the material (type and form of steel, heat treatment condition of steel, etc.).
3. Review of data now available from several reactor surveillance programs and correlation with results from experimental irradiation studies.
4. Analysis of 1, 2, and 3 so as to project the level of neutron induced change relative to some point considered critical to an individual reactor vessel.
5. The summary of the analysis and factors to be considered to assure an optimum surveillance program for future reactors.

This approach requires a thorough analysis of the published data on the engineering aspects of metal fracture, the accepted criteria for assuring operational reliability of steel reactor pressure vessels, the role of radiation in this process, and the condensed results of individual vessel surveillance programs. Additionally, neutron dosimetry and spectral analysis must be utilized in order to correlate meaningfully experimental (test reactor) data with individual vessel surveillance data. Recent developments for minimizing the problem in future reactor vessels through a better understanding of radiation embrittlement and, hence, improvements in steel quality and in reactor design will be reviewed to indicate progress toward the time when little or no surveillance may be required.

In discussing vessel surveillance, the question of steel embrittlement is constantly in the forefront. The term embrittlement is a relative term and is not intended to suggest a severe condition. In fact, radiation embrittlement may be small or large; it is important in either case because of the important function served by the vessel and the economic consequences of overly conservative design and operating practice as a result of uncertainties. With this approach to surveillance and to the changes wrought by radiation, it is hoped that a practical engineering analysis of data from these several very different, yet related areas, will result and be applicable to general questions of reactor surveillance as well as to very specific questions for aiding operational decisions for individual reactor plants.

Fast Fracture of Steels

Background

Ferritic steels such as those used for nuclear power reactor pressure vessels undergo a marked transition from ductile-to-brittle fracture behavior over a temperature range of about 100 to 150 F. (Primary requirements for vessel steel transition characteristics have been established by the ASME Code for Nuclear Vessels)[*1*][1]. The microscopic and macroscopic fracture appearance depicts the change occurring in the steel over this temperature range. These range from crystallographic cleavage (with low energy absorption) at the low end to ductile dimpled rupture or shearing (with high energy absorption) at the high end of the curve. This quality of a steel commonly is known as its notch ductility or impact characteristics and usually is measured by means of the Charpy V-notch impact test (ASTM Specification E 23 for Type A specimens) [*2*]. Through a series of such tests over a temperature range, the ductile-brittle transition temperature (DBTT) curve is formed. The main features of the curve of interest to this analysis are the relative temperature for the rapid rise in energy to fracture and the energy required for ductile fracture, the upper shelf portion of the curve. The features of this curve for an unirradiated and irradiated 8 ⅛-in.-thick plate of A533-B steel, representing a specimen from a pressurized water reactor (PWR) core shell plate, are shown in Fig. 1 [*3*].

The significant point to note is the increase of 190 F in the transition temperature from +10 to +200 F at the arbitrarily selected 30 ft·lb level. The curves marked (*L*) were derived from specimens taken parallel (longitudinal) to the primary plate rolling direction at the one quarter thickness (1/4*T*) location and notch perpendicular to the plate surface.

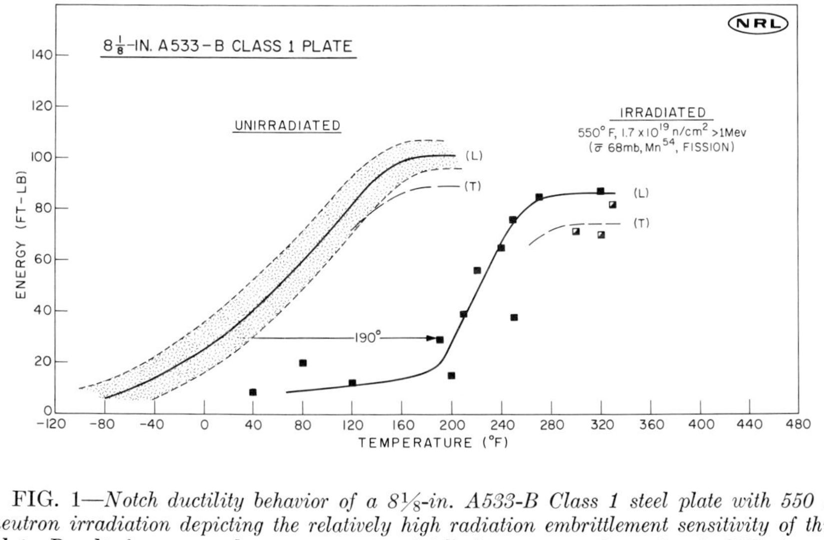

FIG. 1—*Notch ductility behavior of a 8⅛-in. A533-B Class 1 steel plate with 550 F neutron irradiation depicting the relatively high radiation embrittlement sensitivity of this plate. Results from an exploratory assessent of full shear energy absorption in WR (transverse) versus RW (longitudinal) plate orientations also are shown* (*Ref* 3).

[1] The italic numbers in brackets refer to the list of references appended to this chapter

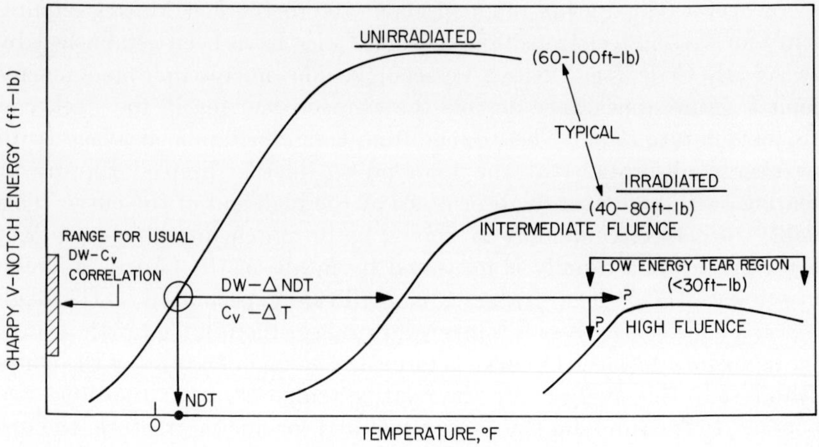

FIG. 2—*Schematic diagram of transition temperature approach to irradiation embrittlement of typical A302-B reactor pressure vessel steels* (*Ref* 4).

Specimens taken in this manner represent the conventional procedure for defining notch ductility characteristics of steel plate since vessel construction orients the plate so that the (*L*) direction represents direction of principal applied stresses. The curves marked (*T*) which for this steel are very similar to the (*L*) curves represent specimens taken across (transverse to) the primary rolling direction. In some plates the variations due to specimen orientation are great enough to warrant concern as will be described later. Similarly, the radiation-induced reduction in energy to fracture at the upper "shelf" of the Charpy V-notch curve for this steel is quite low. Specific and more critical examples of both orientation and "shelf" energy changes will be reviewed later; for reference, these changes are shown schematically in Fig. 2 [*4*].

The left curve of Fig. 2 represents a typical transition in energy absorption (ductile-brittle transition) exhibited by unirradiated notched bar (Charpy V-notch, C_V) impact specimens when tested at progressively higher temperatures. The shelf values are typical for A302-B, later steels of A533-B type exhibit higher energy absorption values. The correlation point of the nil-ductility transition (NDT) temperature on the transition curve varies from steel to steel, but the usual range is indicated by the dashed band at the left. The exact correlation point is important to permit definition of an absolute radiation-induced transition increase, but the 30 ft·lb point provides a typical basis for defining experimental increases. Intermediate fluences (—1 to 2×10^{19} n/cm^2 >1 MeV) at 550 F might result in a 20 to 35 percent drop in the "upper shelf" energy (full shear energy absorption level) and a transition or NDT increase (ΔT) of 100 to 200 F. With very severe neutron exposure, high fluence ($>5 \times 10^{19}$

neutrons (n)/cm^2) or moderate fluence and low (<500 F) temperature, or steels especially sensitive to neutron damage, very great changes as indicated in Fig. 2 may result. If the neutron induced decrease in upper shelf decreases to <30 ft·lb, the possibility of low energy absorption by apparently ductile material results in serious failure potential regardless of temperature.

The most significant facet of radiation exposure for a typical PWR vessel steel, the increase in the DBTT, is well demonstrated by Fig. 1, but the basis for using the shift at the 30 ft·lb level, the measured NDT, is better described below with reference to Fig. 2. The above noted radiation induced changes, and their potential for influencing the fracture characteristics of a reactor vessel provides the basis for vessel surveillance plans.

Nil-Ductility Transition (NDT) Temperature or Energy Absorption Limit

In order to describe a change in transition temperature it is certainly desirable to have a full Charpy V-notch curve for both the unirradiated and irradiated steel. However, if one is to project or describe *a* critical temperature (a temperature below which the stresses on the vessel must be controlled to assure vessel reliability),[2] or *a* critical energy absorption limit,[3] the/full Charpy V-notch curve is inadequate. A single meaningful value must be determined. Herein lies the value of the NDT temperature which when correlated with the C_V data, can be represented by some energy level on the C_V curve. This point is by no means the same for all steels but has been found for several heats of A302-B steel to lie in the range of about 25 to 35 ft·lb, hence the convention of stating ΔNDT or ΔT in terms of the 30 ft·lb value of the C_V. The basis for this is described later.

The NDT temperature concept, through use of the drop-weight test (ASTM E 208–66 T) [*5*], was devised by Pellini [*6*] and co-workers at the

[2] The concept of a critical temperature depends upon selecting a basis for assuring vessel safety through maintenance of the vessel wall temperature above a given temperature while the vessel stresses are above some established level. This critical point will vary with each reactor vessel and is determined primarily by the specific vessel steel and its operating temperature, neutron exposure, and stress conditions. For example, the Army Nuclear Power Program selected a very conservative basis for their SM-1A reactor's critical level [*4*]—a vessel exposure sufficient to raise the vessel NDT temperature +60 to 300 F. If this point were reached, stresses on the vessel were to be limited to 18 percent of yield strength. This basis is conservative but was based on the fact of being the first experience of this sort and that the vessel neutron flux was high, the temperature low, and the steel sensitive to radiation embrittlement.

[3] The concept of *a* critical energy absorption limit on the Charpy V-notch curve and without recourse to the NDT determination has been suggested as an alternative which considers both the increase in transition temperature and the reduction in Charpy V-notch energy absorption with neutron irradiation. The following discussion of the NDT temperature concept and its use is valid, but the concept may be superceded by the Charpy V-notch energy absorption concept now being formulated but not yet fully established.

Naval Research Laboratory (NRL). The broad applicability of the NDT concept has been well demonstrated in the analysis of a number of failures of structures made from the low- and medium-strength steels; that is, less than about 100 ksi yield strength [7]. These analyses related the drop-weight NDT test results with the stress, temperature, and flaw conditions for the failures. The drop-weight test is designed to define the critical temperature transition point for brittle fracture by applying just over yield stress loading (effected by a predetermined weight drop height and "stops" on the anvil) to a specimen containing a small sharp flaw. The change in fracture behavior from "break" to "no-break" is usually quite sharp, occurring over a narrow temperature range (±20 F). Fortunately, the change usually resides on a predicatable low energy portion of the transition temperature range defined by the conventional C_V test (as indicated by the 25 to 35 ft·lb value for A302-B). This is important since the smallest acceptable (valid) drop-weight (DW) test specimen [5] is 5/8-in.-thick by 5-in.-long by 2-in.-wide compared to the 0.394-in.-square by 2.16-in.-long C_V specimen (Fig. 3). Irradiation of the DW test specimen series is difficult. However, the correlation of DW and C_V, as shown schematically in Fig. 2, obviates this problem. Correlation involves determination of the temperature for NDT (break to no-break) and indication of a C_V energy at this temperature (see first curve Fig. 2). Of course, correlation on an unirradiated basis was not equated with correlation irradiated before a number of experiments were conducted and the

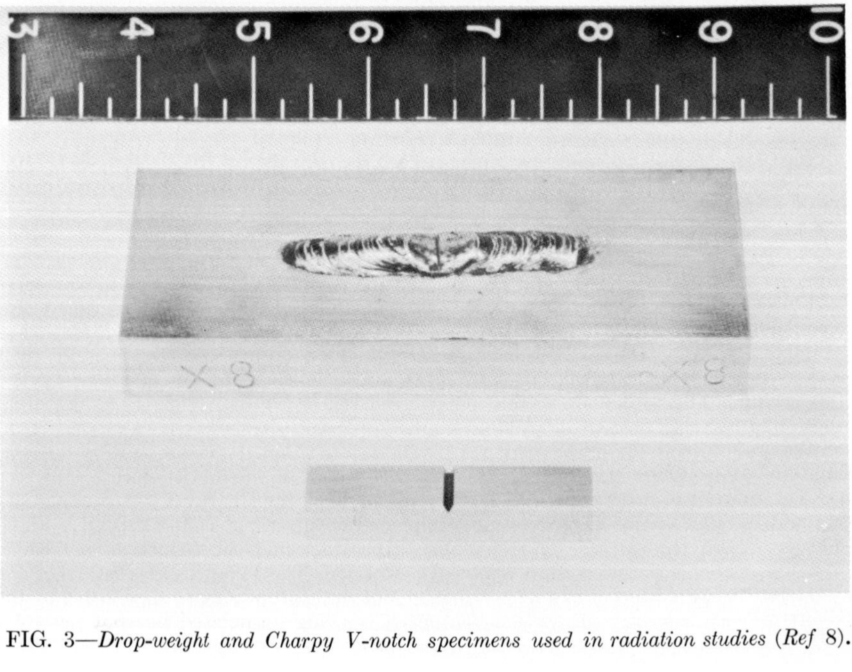

FIG. 3—*Drop-weight and Charpy V-notch specimens used in radiation studies* (*Ref* 8).

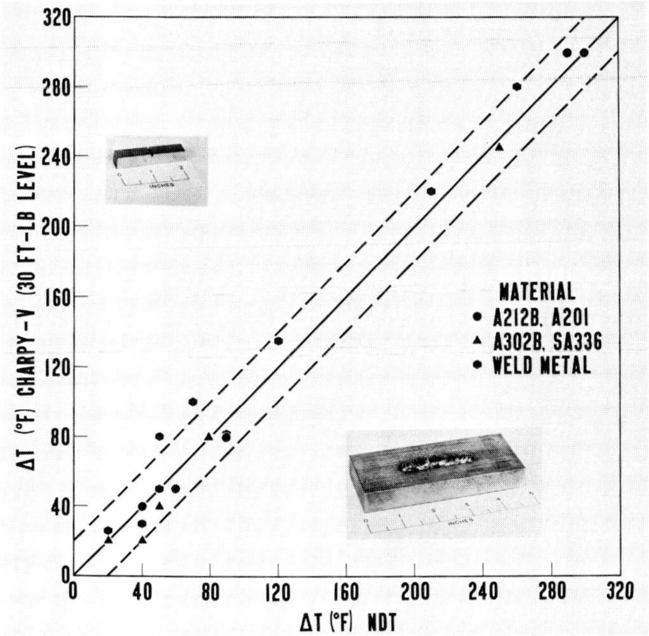

FIG. 4—*Correlation of Charpy V-notch and DW tests of several materials after irradiation* (*Ref* 8).

correlation for A212-B and A302-B steels verified (Fig. 4) [*8*]. This correlation recently has been verified for the A533 steel under conditions of irradiation in the Big Rock Point reactor (BRPR) which very closely simulate the usual boiling water reactor vessel service condition [*9*]. This experiment involved a low irradiation exposure in an accelerated exposure position at a temperature of about 585 F.

NDT and Reactor Vessel Surveillance

Pellini's analysis [*7*] of structural steel failures (nonnuclear) mentioned above shows that they fail by fast "brittle" fracture when the temperature of the steel is below the NDT temperature and other proper conditions—stress and available flaw—are present. By contrast, at temperatures above the transition range ($\geq$60 to 120 F above the NDT), even in the presence of sharp flaws and yield point loading, the steel will not fail by fast fracture.[4]

[4] An exception to the latter statement involves steels (usually of higher strength, 100 ksi or above) which exhibit relatively low energy levels to fracture even at the top of the transition range. This phenomenon of "low energy tearing" [*10*] is not well understood but is related to the inherent lack of ability of the smallest metal ligament to resist separation. This is not a problem with the steels of current nuclear vessel construction unless neutron exposure is very high. Nevertheless, it is important to consider in assessing vessel reliability in nuclear service, especially as the trend toward the use of higher strength steels continue, or if high stresses can be projected for the vessel steel in a direction transverse to the primary rolling direction.

A growing understanding of the brittle fracture of steels has provided guidance for avoiding or at least minimizing the possible problem. First, the ASME Code [*1*] imposed certain fracture limitations[5] upon the properties of steels used in reactor vessel construction. In spite of the fact that notch impact or NDT properties are imposed only for materials chosen for vessel construction under the Code, the broadly accepted concept of the NDT temperature *plus* a temperature factor is the basis for current explicit or implicit programs for avoiding vessel failure through fast fracture even in the irradiated case. (Acceptance of the Charpy V-notch energy absorption limit concept now being considered by the AEC would obviate the need for use of NDT *plus* a factor as Charpy V-notch surveillance data would suffice alone.)

A second important factor is the use of the NDT determination as the basis for examination and correlation of the conditions under which steel structures fail. Pellini [*12*] has developed around the NDT temperature

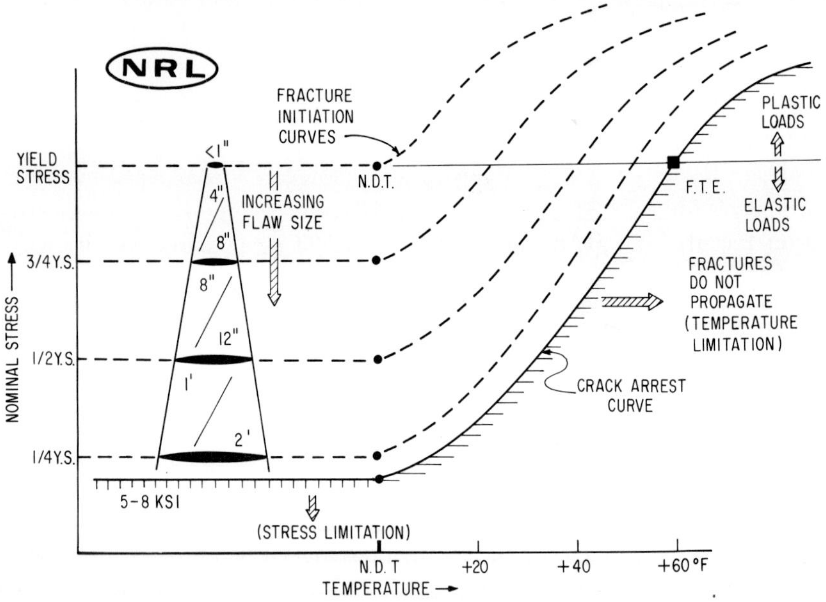

FIG. 5—*Generalized fracture analysis diagram as referenced by the NDT temperature* (*Ref* 12).

[5] Section III of the ASME Boiler and Pressure Vessel Code, Nuclear Vessels (1968) under Paragraph N-331, Ductile-Brittle Transition Tests, requires that vessel steels and other carbon and low alloy steels used in nuclear vessels be tested either by drop-weight or Charpy V-notch impact tests. The NDT temperature is required at a temperature 60 F below the lower of: (1) the vessel hydrostatic test temperature or (2) the lowest service metal temperature. The Charpy V-notch test requirements are specified by type of steel.

a relationship for failed steel structures between the failure temperature, stress, and flaw size. The result is the fracture analysis diagram (FAD) which relates these three factors in a relatively simple diagram (Fig. 5). The diagram applies to steels in the strength range which covers the Type A302-B and A533 steels. It tells us in a semiquantitative way that: (1) with increasing flaw size the stress required for fracture at NDT is reduced progressively or with decreasing flaw size a progressively higher stress is required, (2) with increasing temperature above NDT the stress requirement for fracture, given a flaw, rises rapidly so that, (3) at NDT +60 F (fracture transition elastic) yield stress loading is required to cause fracture, and (4) at the higher temperature of NDT + 120 F (fracture transition plastic) fractures will propagate through plastic deformation (tearing) only by subjecting the steel to its ultimate tensile strength.[6] This development was based upon nonnuclear failures, but, in view of the correlation shown between the drop-weight and Charpy V-notch impact tests before and after irradiation (Figs. 4 and 6), it is reasonable to judge this approach valid for irradiated vessel steels.

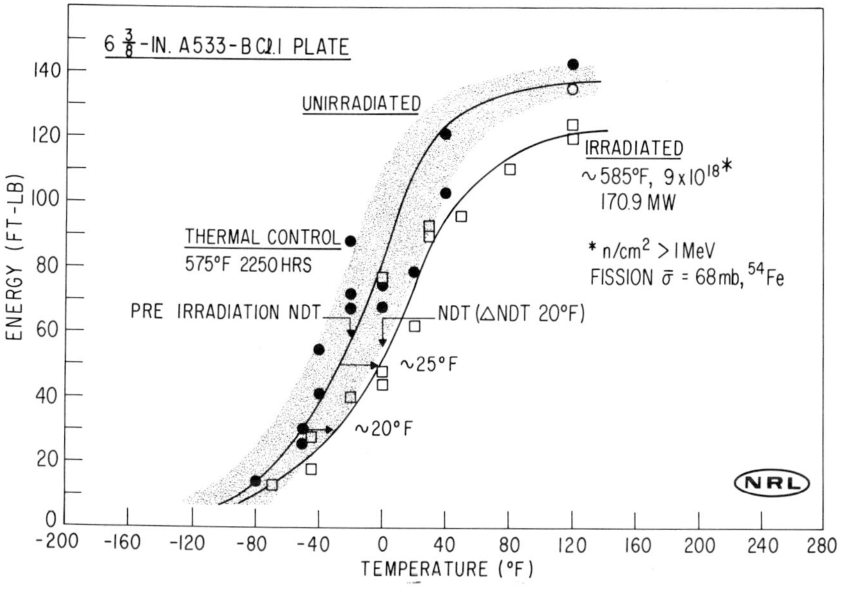

FIG. 6—*Comparison of the Charpy V-notch and DW test performance of 6⅜-in.-thick A533-B Class 1 steel plate irradiated at about 585 F in the Big Rock Point Reactor* (*Ref* 9).

[6] Consult Ref *12* for a more complete explanation and limitations which apply to thin-walled vessels of relatively large diameter. For more recent developments in this subject area see NRL Report 6713 by W. S. Pellini (Ref *20*) and the section later describing the Dynamic Tear Test which involves thicker sections wherein mechanical constraint requires adjustments of the NDT factor for thicknesses greater than about 3 in.

In connection with radiation embrittlement, a number of experimental programs have produced information important to understanding the results of reactor vessel surveillance programs. Using the notch impact specimens mentioned above, the conditions of radiation embrittlement, as a function of the material and the nuclear and physical environment, have been determined for the types and grades of steel of direct interest. While much is yet to be learned about both brittle fracture and radiation embrittlement, meaningful relationships have been established between the two by experimental studies and through the analysis of nonnuclear steel failures. Thus, projection of the results of surveillance programs in terms of what is known today may be invaluable, as these experimental data relationships form the basis for reactor vessel surveillance guidelines published to date.

The ASTM Committee E-10 on Radioisotopes and Radiation Effects has published several standards or recommended practices [*13*] which also relate either directly or indirectly to the subject of surveillance. They include:

Primary:—ASTM E 185, Recommended Practice for Surveillance Tests on Structural Materials in Nuclear Reactors

ASTM E 184, Recommended Practice for Effects of High Energy Radiation on the Mechanical Properties of Metallic Materials

Secondary:—ASTM E 170, Standard Definitions of Terms Relating to Dosimetry

ASTM E 199, Recommended Practice for Measuring Neutron Flux Environment for Reactor Irradiated Specimens

ASTM E 261, Measuring Neutron Flux by Radioactivation Techniques

ASTM E 262, Measuring Thermal-Neutron Flux by Radioactivation Techniques

ASTM E 263, Measuring Fast-Neutron Flux by Radioactivation of Iron

This is not to imply that the background for reactor vessel surveillance is developed fully; it is not. In fact, one of the major objectives of this analytical study is to assess the effectiveness of existing surveillance plans and to make recommendations in terms of developing knowledge in both radiation embrittlement and steel fracture.

Recent Developments in the Study of Steel Fracture

The quest for a more quantitative evaluation of fracture in heavy steels has led to a number of developments in programs which are just beginning to provide conclusive results. Nevertheless, because of their potential importance to the vessel surveillance question, a brief review is provided.

AEC Industry Programs

The growing size and importance of nuclear power and the commensurate need for care relative to safety of the containment system in recent years has focused attention on the need for certain pressure vessel studies. This need was dramatized in late 1965 when the U.S. Atomic Energy Commission's Advisory Committee on Reactor Safeguards (ACRS) addressed a letter on the subject of Reactor Pressure Vessels to the Chairman of the Atomic Energy Commission (AEC). This letter [*14*] addressed the question of safety assurance relative to the primary containment system and recommended, in part, the following:

> To reduce further the already small probability of pressure vessel failure, the Committee suggests that the industry and AEC give still further attention to methods and details of stress analysis, to the development and implementation of improved methods of inspection during fabrication and vessel service life, and to the improvement of means for evaluating the factors that may affect the nil ductility transition temperature and the propagation of flaws during vessel life.

The letter received serious attention from both the nuclear industry and the AEC. The industry response was manifested in a program recommended by the Pressure Vessel Research Committee (PVRC) of the Welding Research Council. In considering the PVRC recommendations, four questions, growing out of the ACRS recommendation, were basic:

1. What are the properties and how effective are inspection techniques for heavy section steels?
2. What effects do flaws and variations of properties have on the behavior of vessels fabricated of heavy section steels?
3. What new or refined inspection techniques may be applied to weldments?
4. What inspection techniques may be applied effectively to vessels in service?

A fifth question, which was at first not treated by the PVRC recommendation, received strong emphasis from the AEC; this was the question of the influence of radiation effects in recognition of operating environmental factors.

As a program for answering these questions began to take shape, three major areas of endeavor were identifiable. These included: (1) materials properties, (2) fracture characteristics, and (3) nondestructive inspection. In recognition of existing programs and inherent interests and capabilities, it generally was agreed that a cooperative effort would involve the AEC's performing the first two tasks and industry the third. (It should be noted that under the auspices of various subcommittees of PVRC, pressure vessel steel properties data are being gathered by reactor manufacturers supplementing the AEC supported efforts on materials properties.) Of course, no absolute division of effort was possible, rather the programs were planned so as to be cooperative and complementary. After considerable review by various U.S. experts, the programs outlined below were accepted with division of primary responsibility as listed.

The broad objective of the AEC and industry programs is to assure structural reliability of reactor pressure vessels constructed of heavy section steels through better understanding of the properties and failure characteristics and the application of improved methods of nondestructive examination during vessel fabrication and during service.

The AEC supported portion of the task for meeting this objective is embodied in the Heavy Section Steel Technology Program and the industry portion in the Industry Cooperative and Industry Funded Programs. The primary content of these programs is outlined.[7]

Industry Cooperative Program (ICP)[8] (carried on through PVRC Subcommittee and Task Groups)—

1. Systemmatically collect and store data on properties and nondestructive examination of plates, welds, and forgings of steels used in reactor vessel construction (now underway).

2. Assess the current status of nondestructive examination capability on heavy section steels through a planned program of cooperative examinations of deliberately flawed, rejected, and accepted (but flawed) plates, welds, forgings, and castings (now underway).

3. Building upon the results of Items 1 and 2: (*a*) provide improvements in resolution, sensitivity, and interpretation of nondestructive examination equipment and techniques, (*b*) develop and evaluate improved methods for examination of heavy section steels, especially weldments, and (*c*) improve the procedures for automatic recording of results of nondestructive examination.

Industry Funded Program (IFP)—

1. Examination of methods having potential for in-service inspection of vessel for indications of the growth of critical flaws. (Initial paper study begun in mid-1967.)

2. Develop and evaluate new methods for inspecting vessels in service (now underway). A major program for meeting these objectives was

[7] Two major program-oriented documents were produced describing needs and objectives of both programs. They were: (1) PVRC Program on Heavy-Section Steel for Presentation to AEC Staff, 8 June 1966, and (2) Memorandum: Reactor Vessel Safety Research and Development Program Based on a Proposal by the Pressure Vessel Research Committee from Milton Shaw, director, Division of Reactor Development Technology to George M. Kavanaugh, chairman, Steering Committee on Reactor Safety Research, 18 Nov. 1966.

[8] Progress in this program was provided by L. J. Chockie and C. F. Larson in a report entitled, "Industry Cooperative Program on Heavy-Section Steels," presented at the 1968 Fall Meeting of the Metallurgical Society of the American Institute of Mining, Metallurgical, and Petroleum Engineers, Detroit, Mich., 16 Oct. 1968.

initiated by the Edison Electric Institute in 1968 under the joint jurisdiction of Southwest Research Institute (SwRI) and the Battelle Memorial Institute-Pacific Northwest Laboratory (B-NW).

AEC Heavy Section Steel Technology Program (HSST)—This effort was authorized by the AEC in March 1967 with the Oak Ridge National Laboratory (ORNL) being designated the co-ordinating organization. The basis and general scope of this effort has been stated as follows [*15*]:

> The Heavy Section Steel Technology Program is an engineering effort aimed at answering specific questions about the structural integrity of light water reactor pressure vessels which contain flaws and material inhomogeneities. These questions will be answered by reevaluating state-of-the-art methods and extending these methods to develop the necessary technologies. First, the information that small standard specimens (with emphasis or transition temperature) reveal about thick plate, vessels fabricated from such plate and actual in-service vessels will be evaluated. Similarly, the applicability of fracture mechanics methods for evaluating vessels under in-service conditions will be investigated. Assuming deficiencies are found, these technologies will be extended to cover the desired temperature and toughness ranges. The program culminates in a series of simulated service tests aimed at validating the total research efforts against pressure vessel tests up to full size and full thickness. Lastly, quantitative engineering methods will be developed which can be described in codes and standards and which can be successfully performed in the average engineering laboratories. The development of the related technologies necessary to make meaningful use of these results and methods will be defined and encouraged.

This program is more specifically outlined in the context of the several objectives [*16*]:

1. Selection, procurement, and characterization of typical and potential heavy section pressure vessel steels (includes PVRC furnished material from vessel construction). Documentation of properties and distribution of this information along with steels to selected subcontracting investigators also is included.

2. Effect of process variables on properties of pressure vessel steels: establish the significance of mill processing variables relative to uniformity and mechanical properties obtained, considering especially:

(*a*) Composition.

(*b*) Heat treatment.

(*c*) Microstructure.

(*d*) Rolling.

(*e*) Thickness effects.

(*f*) Simulation of mechanical treatment of fabrication.

(*g*) For welds—the variations in properties in weld and heat affected zone (HAZ) attributable to welding technique, consideration of alternate techniques and ways to improve welds and HAZ relative to base plates and forgings.

3. Fracture characteristics of heavy section steels, compared to those in current service, and in relation to the expected service environment.

(And in terms of quantitative relationships between flaw size, stress level, and temperature necessary to induce or preclude "low stress" fracture of pressure vessels.)

(*a*) Effect of thickness on fracture parameters, especially energy-absorption and ductile-brittle transition temperature.

(*b*) Correlation of alternate fracture mechanics approaches, assessment of their validity in temperature region near NDT and of their capacity for providing useful quantitative relationships between flaw-size, stress level, and temperature.

Other substudies within the general framework of this task include:

(*a*) Effects of strain rate on notch toughness.

(*b*) Empirical determination of criteria for crack initiation.

(*c*) Determination of fracture toughness values in terms of crack initiation and crack arrest as well.

(*d*) Influence of notch acuity on measured fracture toughness.

(*e*) Phenomena leading to small crack growth to critical size for given stress-temperature condition with emphasis upon fatigue crack propagation.

(*f*) Examination of factors of thermal and strain aging embrittlement of several advanced or potential pressure vessel steels.

4. Effects of radiation on HSST program steels. This effort will parallel the several efforts of the HSST program outlined above adding the factor of the nuclear environment which is pertinent to the range of U.S. water reactor applications—both those in being and those contemplated. Examination of other factors critical to steel failure such as applied stress plus radiation, warm or cold prestressing as a technique to preclude failure, fatigue crack propagation, influence of radiation on energy-to-fracture of steels, influence of radiation and complex stress states, etc. will be studied as well.

5. Correlation of results of fracture characterization studies, (including irradiation effects) and application to service conditions anticipated for heavy section steel reactor vessels and projection of results to the design of reactor vessels so as to optimize structural reliability.

This objective relates essentially to the operating integrity of reactor vessels and involves primarily a critical condensation of the results obtained from programs described previously and the application of these results in the performance of various *simulated service tests* and in the evaluation of the possibilities of *periodic proof testing* through warm (or cold) stressing of a vessel as a means of precluding its untimely failure. This effort also must consider *irradiation effects* and their implication to the structural integrity of a vessel.

The above outlined AEC and industry supported studies should have profound influences upon the understanding of the potential for fracture in reactor pressure vessels, but comprehensive conclusions on the full program are yet several years away. Accordingly, it is necessary that data from current surveillance programs coupled with results of experimental programs be analyzed carefully in order to make meaningful decisions relative to the safe operating practice for operating reactors and to the design of reactors to minimize the problem of vessel embrittlement. Further, great care must be taken to assure that the most productive surveillance programs be installed in future reactors.

In spite of the relative newness and inconclusiveness of the AEC industry studies, certain programs, notably in fracture mechanics and in dynamic fracture testing in thick sections, have made significant progress which is reviewed briefly.

Advances in Fracture Evaluation Techniques

As indicated previously, the predominant consideration in fracture analysis is currently the NDT temperature and, for nuclear vessels, the upward shift in this value for steels exposed to high energy neutrons. The NDT temperature has certain limitations, especially the lack of quantitativeness relative to its definition of the critical conditions for failure, exact stresses, flaw sizes, and temperatures. Thus, the NDT approach must be considered a conservative, semiquantitative limit concept which has gained wide acceptance because of its growth from and emphasis in application upon the actual performance of real structures. Work toward the extension and improvement of this approach is continuing, especially as related to nuclear vessel steels through the supplementary concept—the dynamic tear test. At the same time other approaches which ultimately may provide precise guidelines for fracture safety are being pursued. In fact, many organizations in the United States are seeking more quantitative approaches to the problem of structural failure through brittle fracture of engineering alloys. The most prevalent efforts involve the use of the linear elastic fracture mechanics approach so as to project the critical crack intensity stress value, for a given alloy system containing a specific flaw or flaw size limit.

Fracture Mechanics—The linear elastic fracture mechanics approach involves a quantitative measure of the stress required to fracture (Mode I-bending) a specimen of precise geometry having a fatigued flaw, at temperatures where plane strain conditions (cleavage) exist for low rates of strain. It has been shown to be valid in application to high-strength alloys in structures amenable to complete stress analysis and nondestructive examination which can define rigorously at least a maximum flaw (and its shape) present in the structure. However, in application to the tougher, lower strength steels used in nuclear reactor vessels, significant difficulties have been experienced in obtaining reproducible critical crack intensity stress values from small laboratory specimens except at very low temperatures.

Linear elastic fracture mechanics studies of irradiated steels have produced limited data, and interpretation has been difficult because of questions of the critical factors involved and of the best specimen for determining a valid crack stress intensity factor at instability (K_{Ic})[9] for vessel steels generally having low strength and high toughness. After exposure to high neutron fluences, however, the steels are more amenable to fracture mechanics evaluation as a result of radiation-induced increases in hardness and strength.

Programs for fracture mechanics evaluation of irradiated steels along with a comparative description of the several specimens being utilized have been summarized [*17*].

One study [*18*] provides limited fracture mechanics data on A212-B steel after irradiation to fluences of $>1 \times 10^{20}$ n/cm² >1 MeV at temperatures below 350 F. Another steel (A350-LF3) was irradiated to the same high fluence as the A212-B but at a temperature of 450 F or greater. Charpy V-notch 20 ft·lb results indicated an NDT temperature increase of 615 F (the largest recorded in U.S. studies) for the A350-LF3 and a 390 F increase for the A212-B steel. Parallel changes in K_{Ic} and σ_{ys} were, respectively, a loss of 32 ksi$\sqrt{\text{in.}}$ and an increase of 36 ksi yield strength for the A212-B steels. With no body of similar data available it was difficult to draw meaningful conclusions, but it was possible to conclude under conditions of the test: (1) serious irradiation embrittlement of the steels is evidenced by the drastic reduction of K_{Ic}, (2) the mathematical approach for defining the plastic zone size (a critical parameter in fracture mechanics) does not hold for irradiated steels, (3) while a correlation of K_{Ic} and the strain hardening exponent was possible for the irradiated steel, the extremely small size of the latter casts doubts on the meaning of the correlation, and (4) much work remains to be done before the full meaning of fracture mechanics values for irradiated steels can be understood and applied.

Another application of fracture mechanics to irradiated materials was made in connection with the test of the 2 5/8-in.-thick PM-2A reactor vessel (Fig. 7) [*19*]. In this case a preliminary scale model of a higher strength steel (simulating the PM-2A vessel) failed under a pressure only 6 percent above that stress which was provided by a calculated K_{Ic} value. On the other hand, the complex test sequence for the actual PM-2A vessel yielded a failure K_{Ic} value of 84 ksi$\sqrt{\text{in.}}$ versus a calculated failure K_{Ic} of 58 ksi$\sqrt{\text{in.}}$ This is indicative of the complexities of this approach.

[9] K_{Ic} equals $\sigma(1.21\ \pi(a/Q))^{1/2}$

where:

σ = gross stress, ksi;
a = critical crack depth, in.; and
Q = dimensionless flaw shape (geometry) parameter.

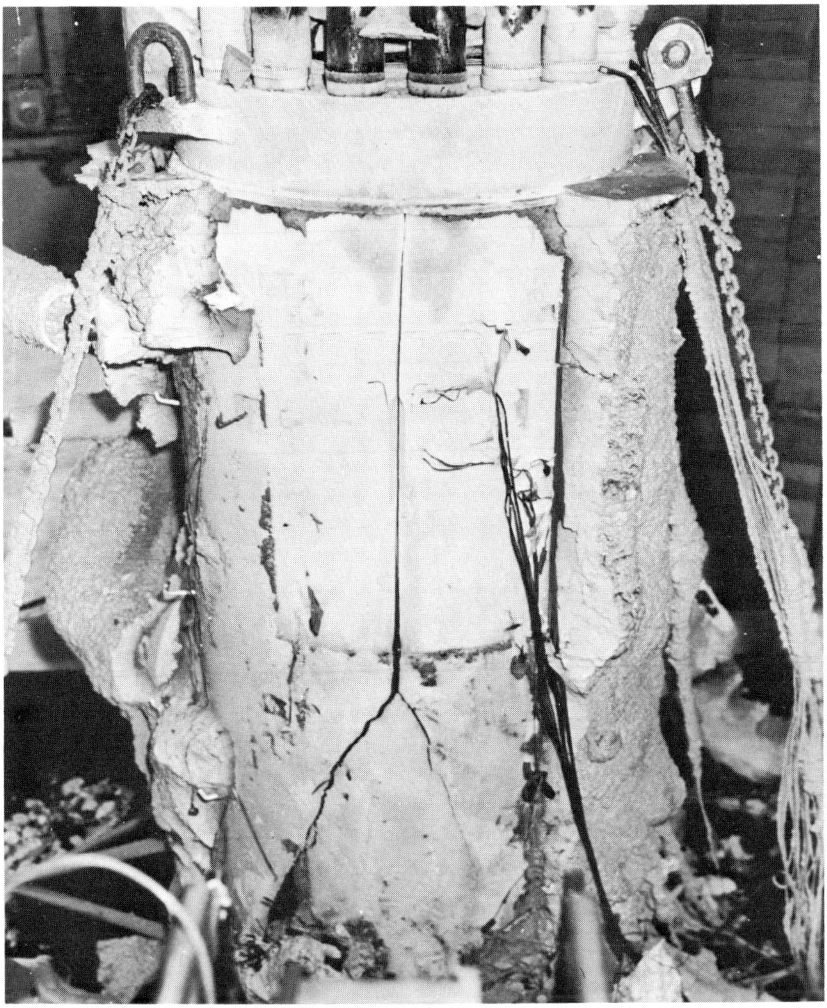

Courtesy of Westinghouse Atomic Power Div.
FIG. 7—*PM-2A vessel immediately after fracture* (*Ref* 19).

In the case of the PM-2A uncertainties in the irradiation level (neutron fluence); in the local applied stress; the uniformity, depth, and sharpness of the stress corrosion crack (Fig. 8) [*19*]; the relative difference in fracture toughness of the corrosion-produced notch versus the usual fatigue crack as well as uncertainties as to the actual wall and cladding dimensions at the flawed location were apparently responsible for a gross miscalculation of failure conditions. Valuable experience has been added through the extension of this study via laboratory evaluations of the PM-2A reactor

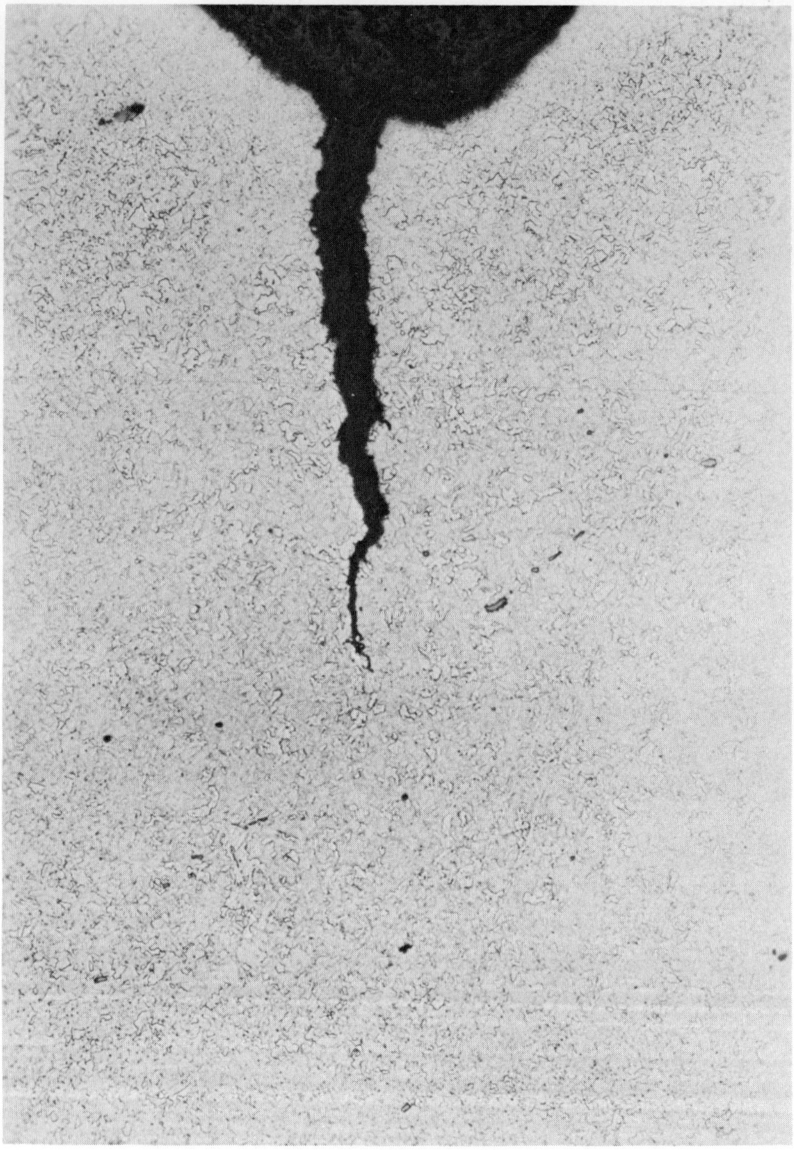

Courtesy of Westinghouse Atomic Power Div.

FIG. 8—*Controlled stress corrosion crack in PM-2A reactor vessel* (*Ref* 19).

vessel steel (see Chapter 3 for further discussion). Nevertheless, this experience is indicative of the criticality of various parameters in fracture mechanics such as flaw size and geometry as well as the geometry and stress distribution in the specimen or in an actual reactor vessel. Other questions relative to the application of fracture mechanics data include:

(1) the meaning of K_{Ic} in terms of yield strength, (2) the effects of dynamic versus static loading (high strain rates have been shown to depress K_{Ic} values), and (3) the validity of very small specimens tested at any low temperature for projecting vessel behavior at service temperature.

The recent work by Pellini [*20*] and his associates demonstrates that the ratio of K_{Ic}/σ_{ys} provides the most meaningful basis for practical application of fracture mechanics. K_{Ic} values alone have little meaning. Recently developed fracture mechanics data on A533-B steels has been provided by the HSST program and is discussed below under "Thickness Effects".

In spite of current limitations to this approach, however, work is continuing at various laboratories and with careful complementary correlation with the more broadly accepted "structural prototype" tests such as the DW and dynamic tear (DT). Herein lies the best hope for quantitative assessment of the fracture safety of nuclear reactor vessels. Recognition of the necessity of this development is implicit in the AEC's supplementary vessel criteria [*14*] which, under the heading "Fracture Mechanics Analyses" reads in part:

> An analysis shall be performed to estimate the margin between the crack size as a result of growth under design cyclic loads and the critical crack size for brittle fracture in the welds of the vessel shell material which directly surrounds the reactor core region. The analysis shall be based on the growth rate of anticipated flaws under design cyclic loads and on material properties at a temperature 60 F above the nil-ductility transition temperature...
>
> The analysis shall demonstate that the estimated fatigue crack size at any time in the vessel service life will be significantly less than the critical crack size for brittle fracture.

Evidence such as this and the growing volume of data on various aspects of fracture mechanics suggests a growing role for this technique as it matures and as the application of higher strength steels becomes more prevalent, thereby aiding the capacity for valid fracture mechanics data. General application for design seems far away but for certain limiting measures as that quoted above, fracture mechanics may have a more immediate and apparent position.

Dynamic Tear Test—Limitations of the small Charpy V-notch specimen for assessing the fracture characteristics become increasingly apparent when the steels considered are the higher alloy steels which are produced in ever heavier sections. Pellini [*10, 20*] and co-workers developed the dynamic tear (DT) specimen to meet the need for an inexpensive and relatively simple test for evaluating the fracture characteristics of higher strength steels and also for the improved heavy section pressure vessel steels. The DT test has been conducted in thicknesses from 5/8 to 12 in., thus providing some critical facts about thickness effects. Further, the nature of the test presents a "limit severity" situation relative to conditions in a structure and provides valuable information over the tempera-

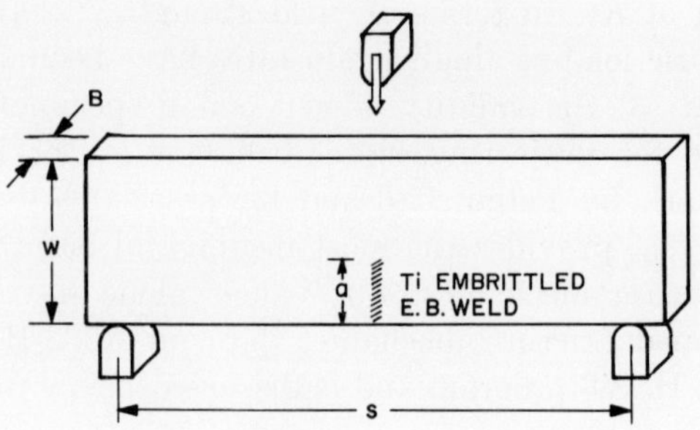

DYNAMIC TEAR TEST SPECIMEN GEOMETRY

Specimen Designation	B	W	S	a	Weight of Steel Spec.
5/8" DT*	5/8"	1-5/8"	6-1/2"	1/2"	2 lb.
1"	1"	4-3/4"	16"	1-3/4"	24 lb.
2"	2"	8"	26"	3"	127 lb.
3"	3"	8"	26"	3"	190 lb.
6"	6"	12"	60"	3"	1220 lb.
12"	12"	15"	84"	3"	4580 lb.

*Alternate 5/8" DT features a 1/2" machined notch terminating in a knife edge pressed tip (5/8" DPN-DT).

FIG. 9—*Features of various DT specimens* (*Ref* 21).

ture range of transition from that of linear elastic to gross strain condition of fracture. The limit severity aspect hinges upon the presence of a large flaw which with the impact (dynamic) loading presents the steel plate with a fast moving crack front across the full plate (or specimen) thickness. The schematic representation of the test and its geometrical variations from 5/8 to 12 in. thickness are shown in Fig. 9 [*21*]. Loss and Pellini [*21*] have described the application of the DT test in relation to linear elastic fracture mechanics evaluations and the changing characteristics of fracture over the next 200 to 300 F which reaches ductile plateau (gross strain condition) for the DT test. A diagram of 5/8-in. DT test results with pictorial views of changing fracture modes over this temperature range is shown in Fig. 10 [*21*]. Fracture appearance changes for the 6 3/8-in. (full thickness) DT test over the temperature range to +160 F (Fig. 11) show the drastic change with temperature. The DT test has been applied over the full range of geometries (5/8 to 12 in.) thick, (Fig. 9) using A533-B plates of 6 3/8 and 12 in. thickness (HSST Plate No. 1). The results for 5/8, 1, 3, and 6 3/8-in. tests of the 6 3/8-in. plate (Fig. 12) are most valuable as they indicate no significant variation in results with increasing specimen size (based upon fracture area), though these represent integration of a variety of steel toughness levels which are indicated

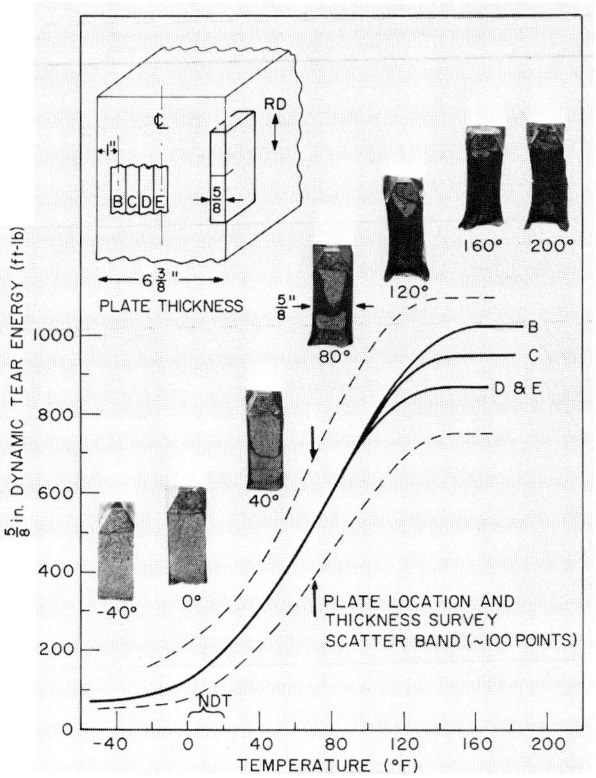

FIG. 10—*Best fit curves of the 5/8-in. DT energy values from four thickness locations of a 6 3/8-in-thick A533-B Class 2 steel plate. The fracture surfaces of selected temperatures are shown. All data are for the RW fracture orientation* (*Ref* 21).

by variations *A* (surface) to *B*, *C*, *D*, and *E* (plate center) except for a remarkable level of energy absorption on the part of the thicker specimens above about midtransition range.[10] The results for two tests of the full 12-in. plate compare favorably with the 3-in. test results. The fracture surfaces for the 12-in. test at 100 F (Fig. 13) indicate rising ductility, most evident in the heavy shear lips.

The results of the DT test for the 5/8 to 12-in. test series have been most encouraging for the reliability of heavy walled reactor vessels. However, because of concern which has been expressed that thick steel sections will fail in a brittle manner at temperatures higher than would be pro-

[10] Even though energy per unit fracture area cannot be expected to correspond, the great energy absorption for the 3-in. test came as a surprise and indicates the fact that gross plastic overload conditions are required for fracture at the level of about NDT + 100 to 150 F.

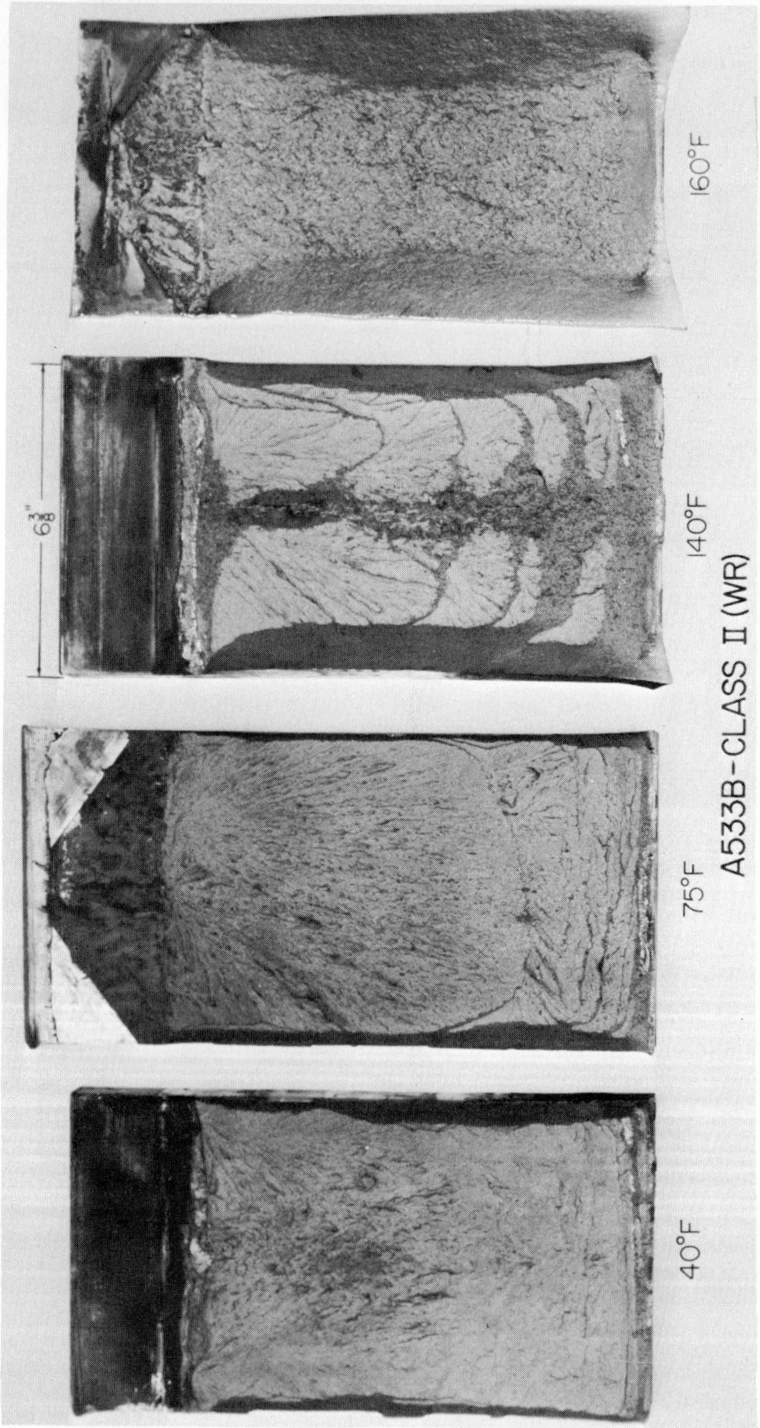

FIG. 11—*Fracture surfaces from a series of full thickness DT specimens from a 6⅜-in.-thick plate of A533-B Class 2 material in the transition region. The specimens are in the WR orientation.*

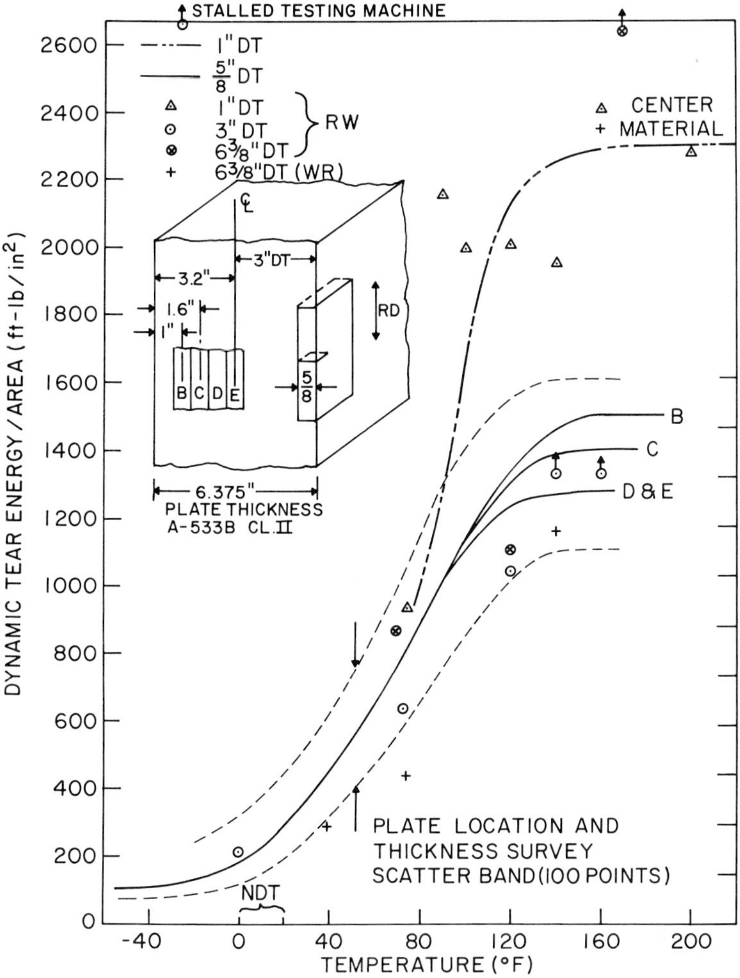

FIG. 12—*A comparison of 5/8, 1, 3, and 6 3/8-in. DT fracture energies from a 6 3/8-in.-thick plate of A53-B Class 2. The fracture energies are normalized by the fractured area* (*Ref* 22).

jected from small (thin) specimens, two critical experiments involving tests (fracture mechanics and dynamic tear) of thicknesses to 12 in. have been conducted. These programs utilized two heats of A533-B steel, one of these from the AEC Heavy Section Steel Technology Program. Westinghouse conducted compact tension specimen tests of 1, 2, 4, 6, 10, and 12 in. thicknesses in order to obtain valid[11] K_{Ic} data at progressively higher

[11] For validity, the ASTM has defined a relative thickness to maximum plastic zone size value of 2.5. With progressively higher temperatures, progressively thicker specimens were required to provide the mechanical constraint necessary to assure compliance with ASTM requirements.

FIG. 13—*Fracture surfaces of both halves of a 12-in. DT specimen tested at 100 F* (*Ref* 22).

temperatures [*23*]. The DT tests were conducted by NRL on specimens of 5/8, 1, 3, 6, and 12 in. thicknesses [*21*, *22*].

Thickness effects—Critical experiments to determine the effects of thickness upon the fracture toughness of thick pressure vessels involved tests over the full transition range.

The results of the DT tests for both 6 and 12-in. plates indicate a sharp transition in fracture toughness evolving for all specimen sizes within a 180 F interval above the NDT temperature [*24*].

Size effects were defined by normalization of DT energy from different size specimens according to the respective upper shelf values. This comparison permitted the definition of a size effect at the 50 percent DT shelf energy level of +70 F between thinner specimens, 5/8 and 1 in., and the 3, 6, and 12-in. specimens. The transition pattern traced by the three thicker specimens is defined by the right curve of Fig. 14. This curve is

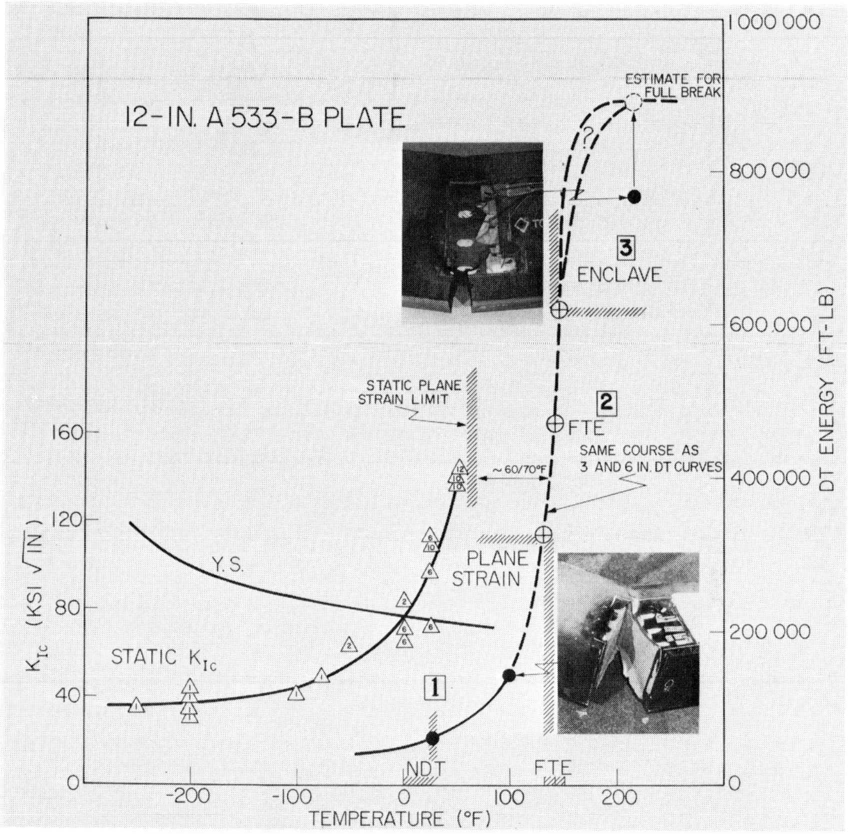

FIG. 14—*Heavy section (to 12-in.) dynamic tear (DT) test results defining a limiting transition temperature for 12-in. A533-B plate steel compared with similar thickness plane strain fracture toughness values* (K_{Ic}). *The sharp rise in* K_{Ic} *at about NDT temperature traces the data from DT specimens of* 5/8 *and 1 in. thickness. Fracture appearance is indicated by photograph inserts. Dynamically tested* K_{Ic} *specimens of the same size would be expected to lower the curve and move it to the right but not beyond the DT curve* (*Ref* 24).

considered a limiting transition temperature in view of the applied dynamic loads to a highly constrained large flaw fracture region. The upper shelf was estimated since it was not possible to completely break the 12-in. specimen with the available 760,000 ft·lb of energy imposed by a falling weight. The transition curve is divided into three sections, the first, the plane strain fracture limit region, the second, the transition region, and the third, the plastic enclave (ductile tearing only) region.

The static K_{Ic} results (up to 12 in.) trace a sharp upward rise at about the NDT temperature. This region of the K_{Ic} curve corresponds approximately to the sharp transition exhibited by the 5/8 and 1-in. DT specimens. Dynamic K_{Id} results are not expected to extend the region of validity for linear elastic fracture mechanics beyond the region defined by the dashed bracket limits of Region 1.

The important conclusion indicated by these studies (Fig. 14) is that a limiting transition temperature, based upon very severe fracture conditions has been defined by DT tests and verified by large K_{Ic} tests. This limit in turn defines a temperature of about 120 to 160 F above NDT above which failure in steels of the quality of A533 could be caused only

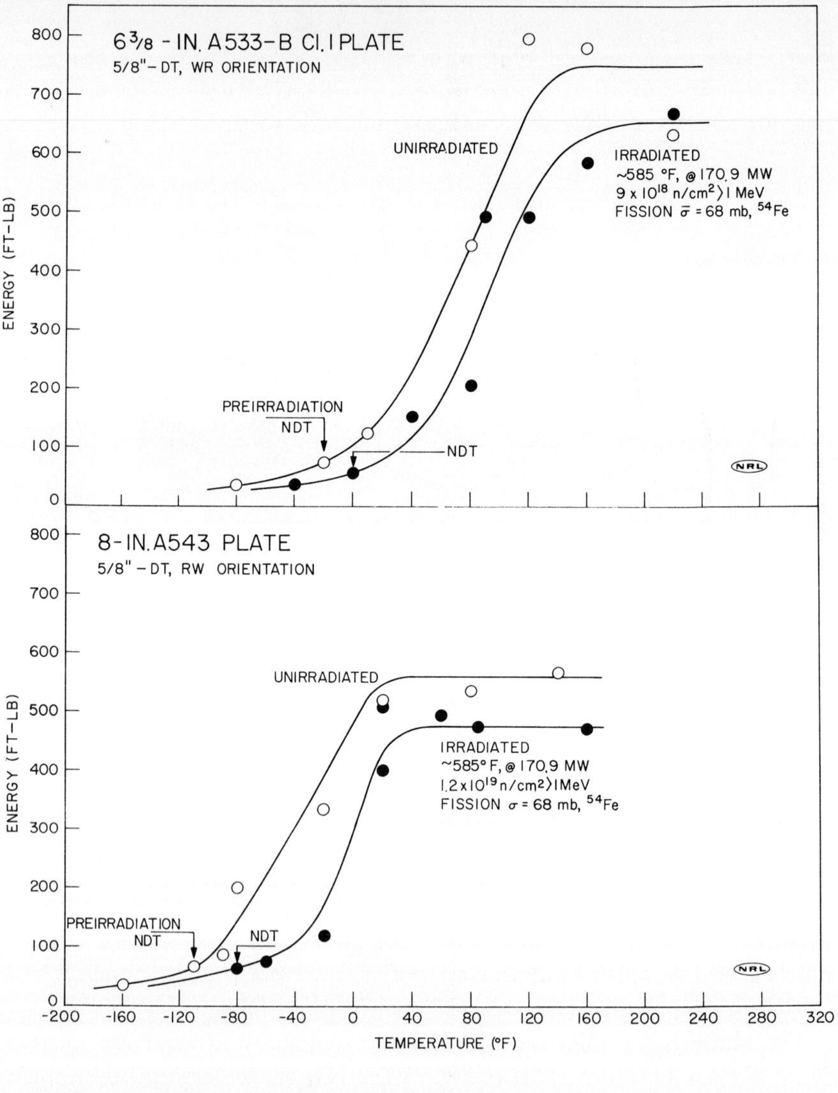

FIG. 15—*DT test energy versus temperature for ⅝-in.-thick specimens of A533-B Class 1 steel and a A543 steel irradiated at 585 F in the Big Rock Point reactor* (*Ref* 25).

by very large flaws and gross plastic loading, conditions which are not expected in operating reactors. This conclusion may be invalidated by severe radiation exposure which might lower the upper energy shelf to a point where plane strain fracture might again occur. The effects of irradiation are being tested using the DT tests.

Dynamic Tear Test Irradiations—Initial irradiations of the 5/8-in. DT specimens have been completed (Fig. 15) in the Big Rock Point reactor (BRPR) and fortunately show little departure from the C_V results obtained in simultaneous irradiations [*25*]. It should be noted that these irradiations were conducted at a temperature of at least 585 F, hence the relatively small shift in transition temperature. (An exact measure of temperature was not possible in BRPR.) Continuation of similar studies and the correlation or irradiation data with full thickness test results are expected to add valuable data to support the state-of-the-art review of Chapter 2.

References

[*1*] ASME Boiler and Pressure Vessel Code, Section III, Nuclear Vessels, American Society of Mechanical Engineers, 1968 edition.

[*2*] "Standard Methods for Notch Bar Impact Testing on Metallic Materials," ASTM E 23-66, *1967 Book of ASTM Standards*, Part 31, pp. 284–300.

[*3*] Hawthorne, J. R. and Potapovs, U., "Intitial Assessments of Notch Ductility Behavior of A533 Pressure Vessel Steel with Neutron Irradiation," NRL Report 6772, Naval Research Laboratory, 22 Nov. 1968; also *Irradiation Effects in Structural Alloys for Thermal and Fast Reactors, ASTM STP 457*, American Society for Testing and Materials, 1969, pp. 113-134.

[*4*] Steele, L. E., Knighton, G. W., and Potapovs, U., "Radiation Embrittlement of Pressure Vessels and Procedures for Limiting This Effect in Power Reactors," *Nuclear Applications*, Vol. 4, April 1968, pp. 230–244.

[*5*] "Tentative Method for Conducting Drop-Weight Test to Determine Nil-Ductility Transition Temperature of Ferritic Steels," ASTM E 208-66 T, *1967 Book of ASTM Standards*, Part 31, pp. 673–691.

[*6*] Puzak, P.P., and Pellini, W. S., "Standard Method for NRL Drop-Weight Test," NRL Report 5831, Naval Research Laboratory, 21 Aug. 1962.

[*7*] Pellini, W. S. et al, "Review of Concepts and Status of Procedures for Fracture-Safe Design of Complex Welded Structures Involving Metals of Low to Ultra-High Strength Levels," NRL Report 6300, Naval Research Laboratory, June 1965.

[*8*] Hawthorne, J. R., "Studies of Radiation Effects and Recovery of Notch Ductility of Pressure Vessel Steels," British Iron and Steel Institute Special Report 69, *Symposium on Steels for Reactor Pressure Circuits*, London, England, 1 Dec. 1960, pp. 343–369.

[*9*] Steele, L. E., et al, "Irradiation Effects on Reactor Structural Materials, Quarterly Progress Report, 1 Aug.–31 Oct. 1968," NRL Memorandum Report 1937, Naval Research Laboratory, 15 Nov. 1968.

[*10*] Pellini, W. S. and Puzak, P. P., "Practical Considerations in Applying Laboratory Fracture Test Criteria to the Fracture-Safe Design of Pressure Vessels," NRL Report 6030, Naval Research Laboratory, 5 Nov. 1963; also *Transactions*, American Society of Mechanical Engineers, Series A, *Journal of Engineering for Power*, Vol. 86, Oct. 1964, pp. 429–443.

[*11*] Steele, L. E., Hawthorne, J. R., and Gray, R. A., Jr., "Neutron Irradiation Embrittlement of Several Higher Strength Steels," NRL Report 6419, Naval Research Laboratory, 7 Sept. 1966; also *The Effects of Radiation on Structural Metals, ASTM STP 426*, American Society for Testing and Materials, 1967, pp. 346–370.

[12] Pellini, W. S., Steele, L. E., and Hawthorne, J. R., "Analysis of Engineering and Basic Research Aspects of Neutron Embrittlement of Steels," NRL Report 5780, Naval Research Laboratory, 17 April 1962; also in *Radiation Damage in Solids*, Vol. II. International Atomic Energy Agency, 1962, pp. 113–151.

[13] "Physical and Mechanical Testing of Metals, Nondestructive Tests," Part 31, and; "General Test Methods," Part 30, *1970 Book of ASTM Standards.*

[14] AEC Press Release No. IN-817 of 25 Aug. 1967, "AEC Develops Regulatory Supplementary Criteria for Nuclear Pressure Vessels," entitled 'Tentative Regulatory Supplementary Criteria for ASME Code-Constructed Nuclear Pressure Vessels.'

[15] Witt, F. J., "Program Plan for Heavy Section Steel Technology Program," 1 April 1968.

[16] Steele, L. E., "The Heavy Section Steel Technology Program," Presentation to Libby-Cockcroft Exchange Meeting, Harwell, England, 27–29 Sept. 1967.

[17] Hawthorne, J. R. and Loss, F. J., "Availability of Data on Irradiated Materials as Related to Design Requirements for Water Cooled Reactor Pressure Vessels," NRL Report 6625, Naval Research Laboratory, 1 Aug. 1967.

[18] Sullivan, A. M., "Fracture Mechanics and Radiation Embrittlement," *Proceedings*, First International Conference on Fracture, Sendai, Japan, 1965, Japanese Society for Strength and Fracture of Metals.

[19] Monahan, J. M. and Walker, T. J., "PM-2A Reactor Vessel Test Description of Testing and Failure Conditions," WAPD-TM-640, Westinghouse Atomic Power Division, Jan. 1967.

[20] Pellini, W. S., "Advances in Fracture Toughness Characterization Procedures and In Quantitative Interpretations to Fracture-Safe Design for Structural Steels," NRL Report 6713, Naval Research Laboratory, 3 April 1968.

[21] Loss, F. J. and Pellini, W. S., "Dynamic Tear Test Definition of the Temperature Transition from Linear Elastic to Gross Strain Fracture Conditions," NRL Report 6787, Naval Research Laboratory, Nov. 1968.

[22] Loss, F. J., "Heavy Section Steel Toughness Characterization, Progress Report Ending 31 August 1968," NRL Memorandum Report 1985, Naval Research Laboratory, March 1969.

[23] Wessel, E. T., "Linear Elastic Fracture Mechanics for Welded Steel Pressure Vessels: Material Properties Considerations," *Symposium on Fracture Toughness Concepts for Weldable Structural Steels*, Culcheth, England, 28–30 April 1969.

[24] Sterne, R. H., Jr., and Steele, L. E., "Steels for Commercial Nuclear Power Reactor Pressure Vessels," *Nuclear Engineering and Design*, Vol. 10, 1969, pp. 259–307.

[25] Serpan, C. Z., Jr., and Watson, H. E., "Fracture Characteristics of A533-B and A543 Steels after 585 F Irradiation in a Power Reactor," Report of NRL Progress, Naval Research Laboratory, Dec. 1968, pp. 18–19.

CHAPTER 2—NEUTRON EMBRITTLEMENT OF PRESSURE VESSEL STEELS—A BRIEF REVIEW

Introduction	49
Environmental Factors Affecting Steel Embrittlement	49
Specimens and Related Factors for Measuring Embrittlement	50
Neutron Fluence	53
Neutron Flux	56
Irradiation Temperature	59
Temperature Effects—Annealing	61
Applied Stress	65
Hydrogen Embrittlement	65
Materials Factors Affecting Steel Embrittlement	67
Variable Embrittlement Response Among Steels	67
Heat-to-Heat Variations	69
Microstructure as a Variable in Radiation Embrittlement Sensitivity	71
Variations Between Plate, Weld, and Weld Heat Affected Zone	74
Initial Results on A533 Steel Plate and Welds	77
Developing Steels for Low Radiation Embrittlement Sensitivity	79
Other Mechanical Properties	90
Tensile Properties	90
Fatigue	97
Summary	99
References	99

CHAPTER 2—Neutron Embrittlement of Pressure Vessel Steels—A Brief Review

Introduction

Nuclear radiation has been shown to affect the properties of structural steels; primarily through hardening and embrittlement. Such changes may be small or large, but in any case they must be defined so that the condition of the reactor vessel is known at every phase of reactor life. This is the basis for reactor vessel surveillance.

A number of experimental programs have approached this problem in the past, and current efforts supported by reactor vessel surveillance programs provide a good basis for understanding the important factors in assessing conditions of a reactor vessel. Thus, a critical review of the current state-of-the-art knowledge of neutron radiation embrittlement is presented as an introduction to and basis for projecting available surveillance data.

Environmental Factors Affecting Steel Embrittlement

Data from various experimental programs conducted over the past ten to twelve years have provided the basis for judging the influence of several important variable factors which can be defined for each reactor pressure vessel. The more important of these factors are: (1) the neutron fluence,[1] (2) the irradiation temperature, and (3) the steel microstructure and composition.[2] Other factors such as steel condition (plate, weld, heat affected zone, forging, etc.), stress state, etc., must be considered as well, but the three listed above predominate.

[1] Neutron fluence cannot be abstracted from other factors which make up the vessel's nuclear environment. Other factors, the distribution of neutrons by energy (spectrum) and the number which move through a given volume in each second (flux), are also important. These factors are considered in detail in Chapter 5.

[2] Certain residual elements, especially copper, phosphorus, and vanadium, have been shown, empirically, to affect drastically the sensitivity of steel to radiation embrittlement.

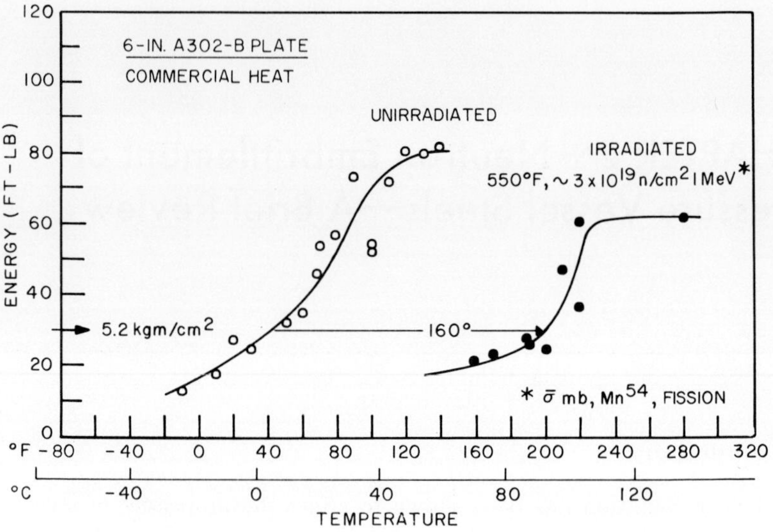

FIG. 16—*Typical Charpy V-notch impact data for irradiated A302-B steel.*

Specimens and Related Factors for Measuring Embrittlement

In studying embrittlement, the Charpy V-notch (C_V) or other specimens for measuring notch ductility are exposed in test, research, or power reactors so as to simulate the exposure at the vessel of a power reactor but usually on an accelerated basis (years of reactor vessel life in weeks or at most months). The C_V specimens usually are taken from the one quarter thickness (¼*T*) level of the steel plate with the long dimension of the specimen oriented in the direction of primary plate rolling and notched perpendicular to the plate surface. Drop-weight specimens also are taken usually from the ¼*T* location though positioning is not so critical. Specimen orientation relative to the primary plate rolling direction may affect the notch toughness properties very significantly. Specimens oriented longitudinally with the primary direction of rolling, the strong (RW) orientation, exhibit higher energy absorption than specimens in the transverse, weak (WR), orientation.[3]

A series of C_V specimens are required to establish a transition curve. Drop-weight specimens (usually six to eight) are used to determine the

[3] In steelmaking, a heavy billet or ingot is rolled repeatedly (longitudinal-strong) and cross rolled (transverse-weak) to reduce thickness and provide a preferred microstructural orientation. The impact specimen is taken usually in the direction of primary rolling (though recently transverse orientation is receiving greater emphasis), and the fracture occurs across the structure that is elongated and strengthened by the rolling action. When specimens are taken in the cross-rolling (secondary-rolling) direction, the fracture occurs in the direction of primary rolling between elongated grain and thereby requires lower fracture energies.

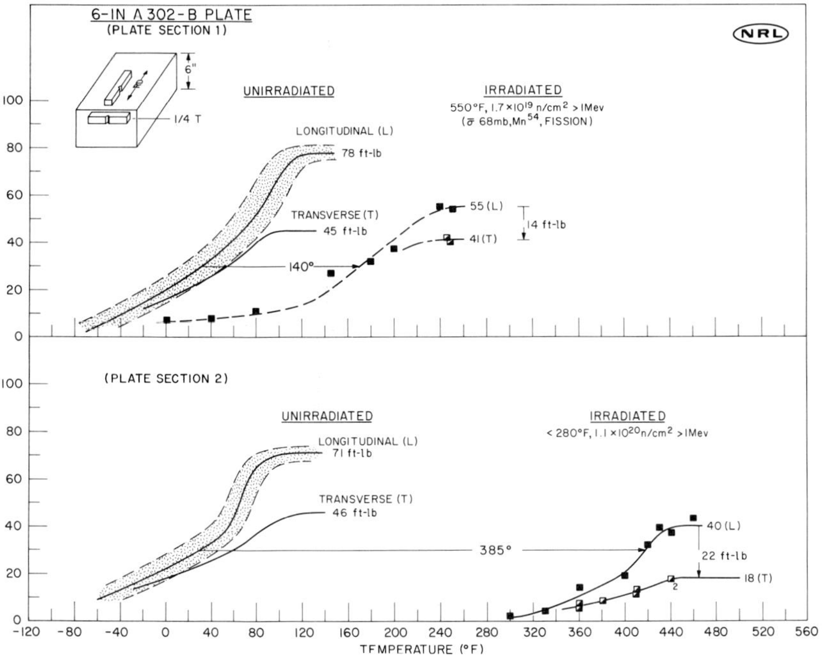

FIG. 17—*Charpy V-notch ductility of ASTM reference A302-B steel plate (specimens taken in longitudinal and transverse orientations to plate rolling direction) after irradiation exposure at 550 F and at 280 F (Ref 2).*

NDT temperature of change from "no-break to break" behavior. By careful correlation (before and after irradiation), an energy (ft·lb) value on the C_V curve is established, and the increase in NDT thus may be determined using the C_V data only. This increase, however, tells only part of the story of radiation effects upon notch ductility. The missing facet from ΔNDT or $\Delta T\text{–}C_V$ data is the sometimes severe reduction of upper shelf, ductile tear energy value, in irradiated steels. These relationships for a typical A302–B steel after intermediate (1 to 5 × 10^{19}) and high (5 to 10 × 10^{19} n/cm^2 > 1 MeV) neutron fluences are shown schematically in Fig. 2, Chapter 1. Typical results for irradiated C_V specimens of the ASTM A302–B reference steel[4] (a 6-in. plate) are shown in Figs. 16 [*1*][5] and 17 [*2*]. These results are very similar to those observed for A533–B steel, though the latter usually

[4] U.S. Steel, about 1957, produced a 6-in. plate of A302-B steel and made it available through ASTM for reference irradiation study. Many experimenters in the United States and other countries have used this steel as a standard for comparison of other irradiated steels. Unfortunately, this material essentially is depleted, though some surveillance programs (those initiated by Westinghouse) have included this steel. Thus, a basis for a very meaningful comparison is provided by this steel.

[5] The italic numbers in brackets refer to the list of references appended to this chapter.

exhibits a higher shelf energy. The relative radiation induced increase in transition temperature and reduction in full shear fracture energy for intermediate fluence irradiation at an elevated temperature (550 F) are compared with similar results for a high fluence-low temperature (<280 F) exposure. A potentially critical factor with this and other steels is the low energy absorption for both strong and weak orientations under severe exposure conditions. Experimental studies of higher strength steels [3] suggest that when the C_V ductile shelf reaches only 30 ft·lb or less, fast fracture may occur by a low energy tearing process, and higher temperatures provide no added deterrent to fracture. This factor must be considered as carefully as the radiation-induced increase in the transition temperature (ΔT) though the point of most concern is the transverse plate orientation which in the vessel would be oriented in a direction other than that of the principal vessel stresses.

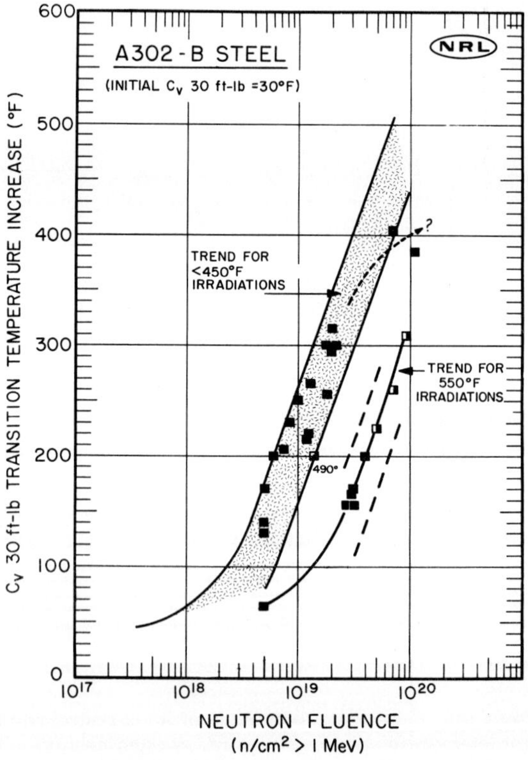

FIG. 18—*Trend of radiation induced transition temperature increase for ASTM 6-in. A302-B reference steel. The half-open symbols represent data obtained from irradiation in the Yankee atomic power plant accelerated exposure surveillance capsule* (*Ref* 18).

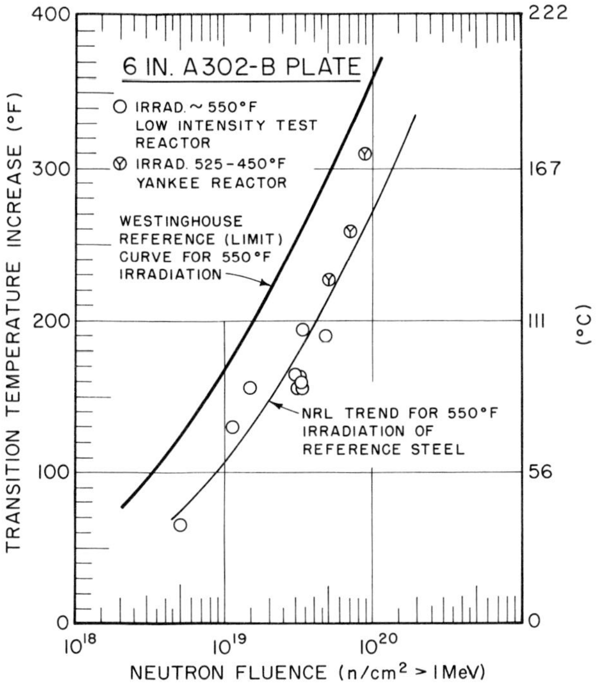

FIG. 19—*Trend of radiation induced increase in transition temperature for A302-B steel irradiated at 550 F with accepted Westinghouse embrittlement trend for 500 F design* (*Ref* 5, 20).

Neutron Fluence

The usual convention for reporting neutron exposure or fluence is in terms of neutrons per square centimeter (n/cm^2) having energies greater than one million electron volts (>1 MeV). The question of damage caused by neutrons having lower energies has not been ignored, and as more is learned about the mechanisms of radiation damage a better approach to defining the damaging neutron fluence (damage as function of neutron energy) is expected. (See Chapter 5 for further discussion of this point.)

Figure 17 illustrates both the fluence and temperature effects. While no data for the 6-in. reference plate are available for elevated temperature irradiation at 1×10^{20} n/cm^2 or greater, the Yankee–Rowe accelerated surveillance data do provide a basis for comparing fluence effects on this one steel. The ΔT for the Yankee irradiation at 9×10^{19} $n/cm^2 > 1$ MeV is 310 F. The Yankee irradiation temperature was apparently only about 525 F with a gradual reduction in the latter one third of each fuel cycle so that the data are not directly comparable with the data in the top part of Fig. 17 but are, nevertheless, quite close. A comparison based upon many irradiation experiments and many fluence levels (including the data from the Yankee accelerated surveillance capsules) is shown in Fig. 18 [*4*].

The fluence effect at two temperature levels <450 F and about 550 F are shown here on a semilog scale. The band shown at left represents the ΔNDT trend for many steels irradiated at temperatures below 450 F where temperature influence is not great. The dashed lines about the 550 F trend are based upon the variations in sensitivity observed between several heats of A302–B steel. The dashed curve at the upper part of the <450 F trend suggests the possibility of embrittlement saturation though the data are too limited to demonstrate this phenomenon conclusively. In Fig. 19 all available 550 F irradiation data for the reference steel are plotted relative to the Westinghouse 550 F design basis curve [*5*]. The Westinghouse curve is apparently very conservative relative to ASTM reference steel. It should be noted that both Babcock and Wilcox [*6*] and Combustion Engineering [*7*] have similar conservative curves. General Electric [*8*] uses individual bases for the older (operating) plants and a similar conservative design curve for newer plants (not yet operating).

A more precise indication of neutron fluence effects in the absence of temperature or materials effects is presented in linear relationship in Fig. 20 [*9*]. The data are based upon irradiation experiments on the heat of A302–B steel as presented in the trend band of Fig. 18 with the irradiation temperature limited to <250 F and including data (low fluence range) from an experiment for simulating the conditions of neutron embrittle-

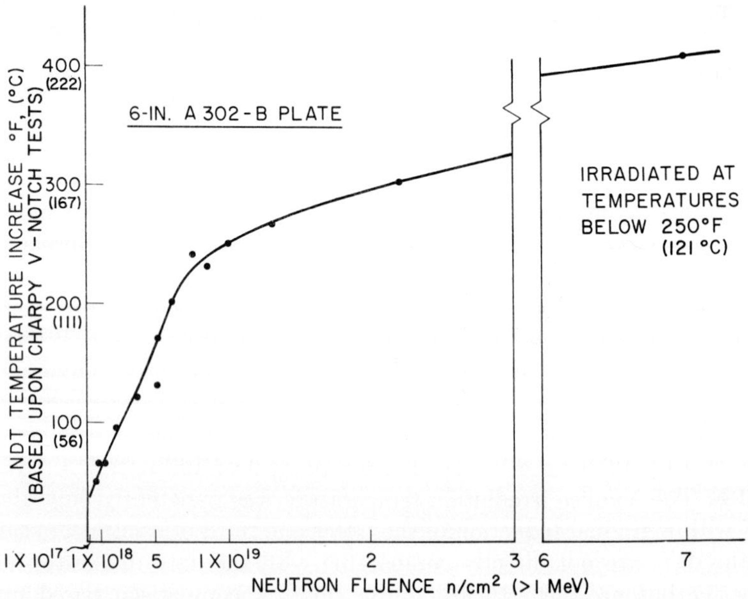

FIG. 20—*Increase in NDT temperature of A302-B steel irradiated to various neutron fluences at temperatures below 250 F.*

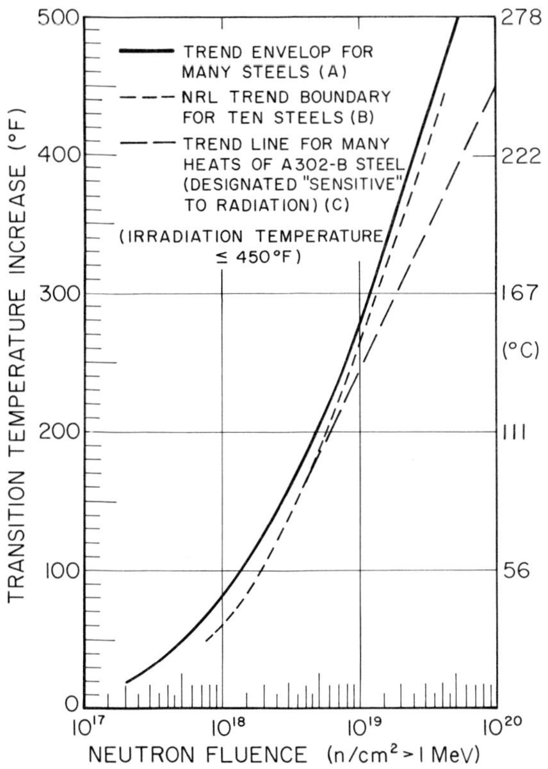

FIG. 21—*Extremes of radiation embrittlement trends based upon three independent surveys of data.* (*Curve* A *from Ref* 11, *Curve* B *from Ref* 12, *and Curve* C *from Ref* 13).

ment through the wall of a reactor vessel [*10*]. The rapid embrittlement to about 1×10^{19} and subsequent indications of saturation ($\Delta T \sim 350$ F at 3.5×10^{19}, $\Delta T \sim 400$ F at 7×10^{19}) have important implications to the application of embrittlement results.

The curves of Fig. 21 represent three trend limits [*11, 12, 13*], the most severe [*11*] being the embrittlement extreme for essentially all recorded transition temperature data and, the least severe, the trend (fitted curve) for so-called "sensitive" heats of A302–B steel as reported by Carpenter et al [*13*]. The close correspondence of these limit trends considering the large number of steels involved is surprising and encouraging. However, it should be noted that these represent the extremes of embrittlement sensitivity, and much data scatter to the right of these limits is experienced when data from many steels are compared. Nevertheless, the good agreement, even for the most severely embrittled steels, suggests that for low irradiation temperatures the trends of embrittlement depend primarily upon neutron fluence with little effect of composition and micro-

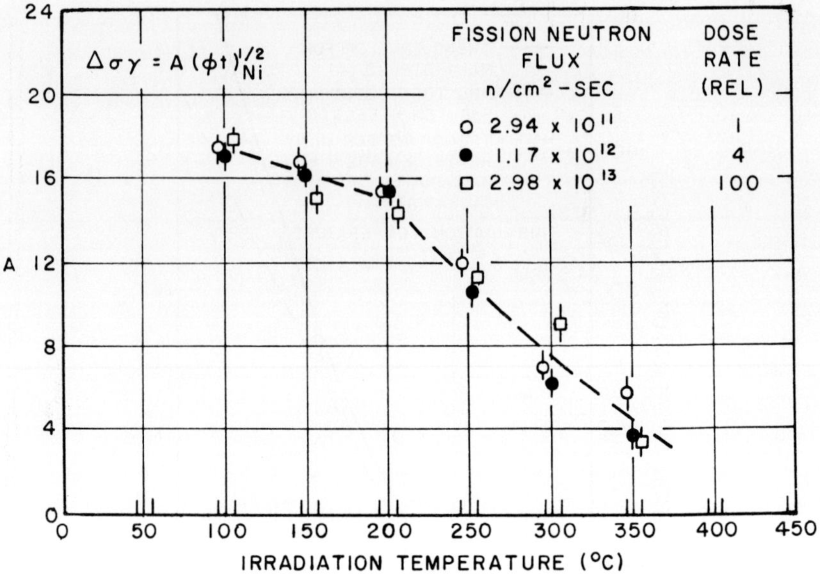

FIG. 22—*Effect of neutron dose rate on radiation hardening of EN-2 mild steel* (*Ref* 16).

structure. The subtle influences of composition, of neutron spectrum variations, and of irradiation temperature, nevertheless, may be quite important as we consider data away from these extremes.

Neutron Flux

Most experimental irradiation studies are conducted in test reactors so as to simulate many years of reactor operation in the space of weeks or at most months. Questions have been raised many times by experimenters as to the validity of such accelerated results.

A theoretical analysis by Gray [*14*] suggested that a high flux rate might provide misleading data. However, two experiments, one by Hinkle et al [*15*] and one by Harries et al [*16*], have demonstrated, respectively, for yield strength increase ($\Delta\sigma_{ys}$) in Ferrovac E iron and $\Delta\sigma_{ys}$ for EN–2 (carbon, silver, manganese) mild steel that there is no flux rate effect for the conditions studied. In both cases the range of flux rates was quite similar being approximately 1×10^{11} to 1×10^{13}. The data produced by Harries et al have more direct interest for this review because the steel used is closer to the question of vessel surveillance. Furthermore, this study covered a range of temperatures, thus providing additional information of value.

Harries and co-workers irradiated tension specimens of EN–2 mild steel in a fuel position of the Pluto reactor and operated the reactor at three different power levels for varying time periods so that the fluence was essentially the same, but the rate varied relatively as 1:4:100 and temperatures ranged from 100 to 350 C (212 to 662 F). The results are plotted in Fig. 22 [*16*]. The values of A are the appropriate factors to describe the temperature effect upon the yield stress for a given steel exposed in a specific neutron spectrum where the relationship $\Delta\sigma_{ys} = A(\phi t)$ is assumed to hold; where A is a constant for the condition and ϕt is the time (second) integrated flux based upon the nickel activation dosimeter in an assumed fission spectrum. The good agreement for the three rates of irradiation are indicative of no flux rate effect.

It should be noted, however, that in both of these experiments, the property measured, $\Delta\sigma_{ys}$, is quite sensitive to irradiation, especially at low irradiation temperatures, that the materials studied are not directly comparable, that the flux rates are still a great deal higher than that expected in large power reactors, and, probably most important, the fluences are quite low (little ΔNDT would be expected), especially in Harries experi-

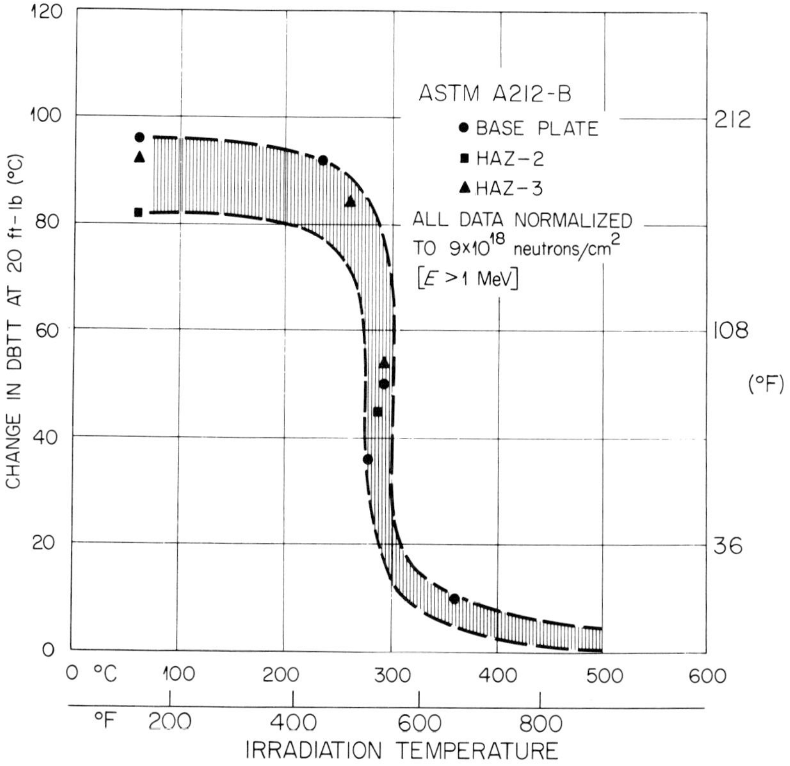

FIG. 23—*Change in transition temperature of A212-B steel after irradiation at various temperatures* (*Ref* 11).

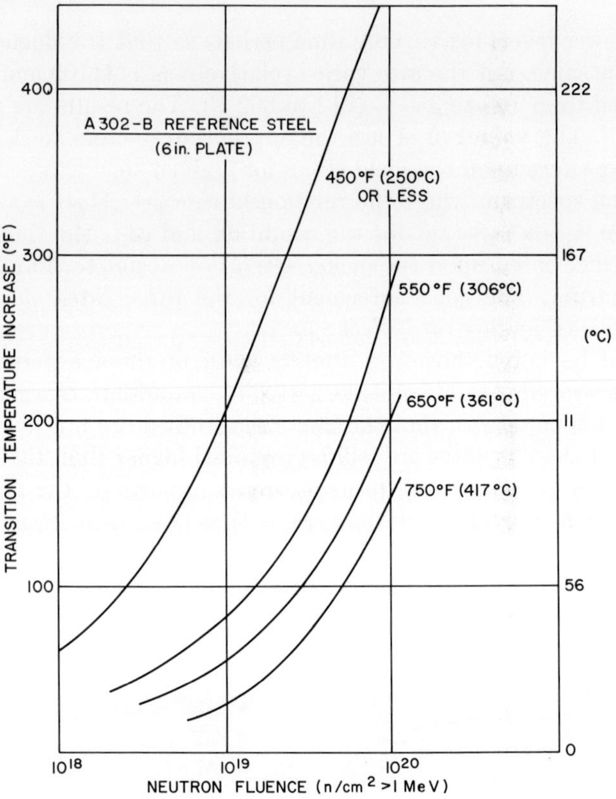

FIG. 24—*Trends of transition temperature increase versus neutron fluence for 6-in. A302-B reference steel irradiated at various temperatures.*

ment (2×10^{17} n/cm^2). Further, the temperature in the Hinkle experiment was not representative of reactor operation being only 95 C (200 F) Thus, while the experimental evidence suggests that there is no flux effect this requires further testing. This is one of the "fringe benefits" to be gathered from reactor surveillance programs. For example, recent tensile data [*17*] from the Big Rock Point reactor surveillance program (accelerated and vessel wall positions) show a distinct effect of irradiation rate on tensile properties but not on notch impact data. Also, flux rate may explain the apparent discrepancy in the level of ΔT shown from the Yankee accelerated versus vessel wall results, the latter showing an apparently more severe embrittlement considering the temperatures and fluences involved (see Chapter 4 for details). However, in the Yankee case, the unresolved question of exact irradiation temperature may enter the picture as well.

The whole question of possible flux effects is largely academic relative to an individual reactor having a good surveillance program since this factor will be "built-in" to the program. Nevertheless, it is a point to consider for those cases where the surveillance program is limited.

Irradiation Temperature

This factor can be most important in assessing and projecting radiation embrittlement in a reactor vessel since with progressively higher irradiation temperatures, progressively lower embrittlement is observed. Note the effect shown in Fig. 23 [*11*]. Thus, intentional application of higher temperature to a vessel may result in ductility restoration through correction ("annealing") of damage.

This temperature effect apparently is based upon the greater mobility of radiation produced defects at elevated temperatures and the related chance for restoring the preirradiation lattice order or their movement to available defect sites at grain boundaries where their effect is to enhance ductility.

The general pattern of irradiation effects with progressively higher temperatures is shown in Fig. 24. This series of trends is based upon a number of irradiation experiments involving the ASTM reference A302–B steel. At 650 and 750 F the curves are largely schematic but are indicative of results at these higher temperatures. A pattern for comparison of the temperature effect on a number of different steels (all U.S. data to 1967) is shown in Fig. 25 [*11*]. It is clear from this figure that different steels respond quite differently to irradiation at different temperatures. This also is shown well by Nichols and Harries [*18*] in a tabulation of data for

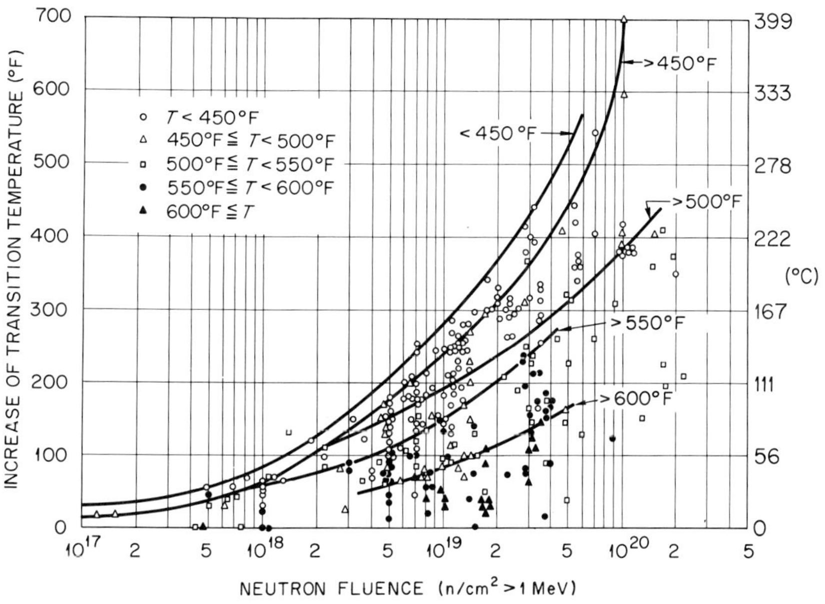

FIG. 25—*Effect of irradiation temperature on the increase in Charpy V-notch transition temperature* (*Ref* 11).

TABLE I—*Effect of irradiation temperature on the radiation induced increases in slow-bend transition temperatures of steels and weld metal* (*Ref* 18).

Integrated Neutron Dose, neutrons per square cm (fission)	Irradiation Temperature, deg C	(deg F)	Increase in Transition Temperature (50% fibrous fracture), deg C								
			Steel 1		Steel 2		Steel 3		Steel 4		Steel 5
			Parent Material	Weld Metal	Parent Material	Weld Metal	Parent Material	Weld Metal	Parent Material	Weld Metal	Parent Material
4.0×10^{18}..........	50	(122)	23	64	31	33	60	...	...	...	...
	130	(266)	53	72	53	40	70	...	...	...	...
	150	(302)	...	...	...	...	...	41	70	43	80
	190	(374)	...	...	...	...	...	...	...	...	71
	245	(473)	...	...	...	...	59	18	42	49	...
	300	(572)	20	30	41	59	62	8	25	12	20
	325	(617)	0	20	...	...	63	8	27	0	0
1.0×10^{19}..........	50	(122)	58	102	58	82	93	...	...	...	...
	130	(266)	72	115	84	101	110	...	...	...	...
	150	(302)	...	...	...	...	...	64	120	108	125
	190	(374)	...	...	...	...	...	...	...	...	112
	245	(473)	...	...	...	...	77	28	66	66	...
	300	(572)	31	47	64	79	97	12	40	18	31
	325	(617)	0	31	...	...	100	12	44	0	0

five steels and related welds irradiated at temperatures between 90 and 585 F (Table 1).

Some recent data [*19*] from an irradiation conducted in the BRPR show vividly how the temperature effect accelerates above 550 F. In this experiment the reference A302–B steel was exposed along with seven higher strength steels. Although no thermocouples were present the BRPR coolant was known to be about 585 F and even if gamma heating added a few degrees, the implications of the data are unchanged. After exposure to 7.1×10^{19}, the ΔT was only 165 F, suggesting that the slope of the higher temperature trend lines of Figs 24 and 25 might be somewhat lower. At 550 F this fluence (7.1×10^{19}) would be expected to produce an increase of about 225 F.

This strong influence of temperature has been utilized in annealing the SM–1A vessel, and studies indicate that such techniques might correct a significant portion of the damage in some irradiated steels.

Temperature Effects—Annealing

The engineering impetus for the study of annealing of radiation embrittled steels came from the compact Army reactors, especially the SM–1A, and much of the available data on this subject relate to that reactor.

It has been observed that several factors influence the level of embrittlement correction obtained by annealing. These include the type of steel (its tendency to embrittlement sensitivity), the temperature of irradiation, and the conditions of annealing (temperature and duration). In general, however, a striking parallel between annealing results and that of irradiation at a specific temperature has been observed. For example, an experiment [*9*] to test the effects of irradiation temperature involved irradiation of a A302–B steel at 400 F to a fluence of 2.3×10^{19} n/cm^2 followed by irradiation at an elevated temperature of 540 F for an added fluence of 1.6×10^{19} n/cm^2. The result was an increase in transition temperature falling just on the 550 F irradiation trend line of Fig. 24.

The effects of annealing time and temperature on the SM–1A steel (A350–LF1) are shown in Fig. 26 [*9*]. Similar patterns have been observed for other steels but generally on a reduced scale because of the greater sensitivity to irradiation (and annealing) of the A350–LF1 steel. A detailed review summary of SM–1A vessel embrittlement data and of the subsequent successful annealing operation is provided in Chapter 3. The high sensitivity of the SM–1A vessel steel, its low operating temperature, and high operational flux all work to make the SM–1A case unique and demonstrative. These same factors, fortunately, work in the opposite direction for the newer and larger reactors.

One factor in operation of some reactors may work to offset this advantage however. This is the tendency to allow the operating temperature of a reactor to tail-off near the end of a fuel batch in order to extend its life.

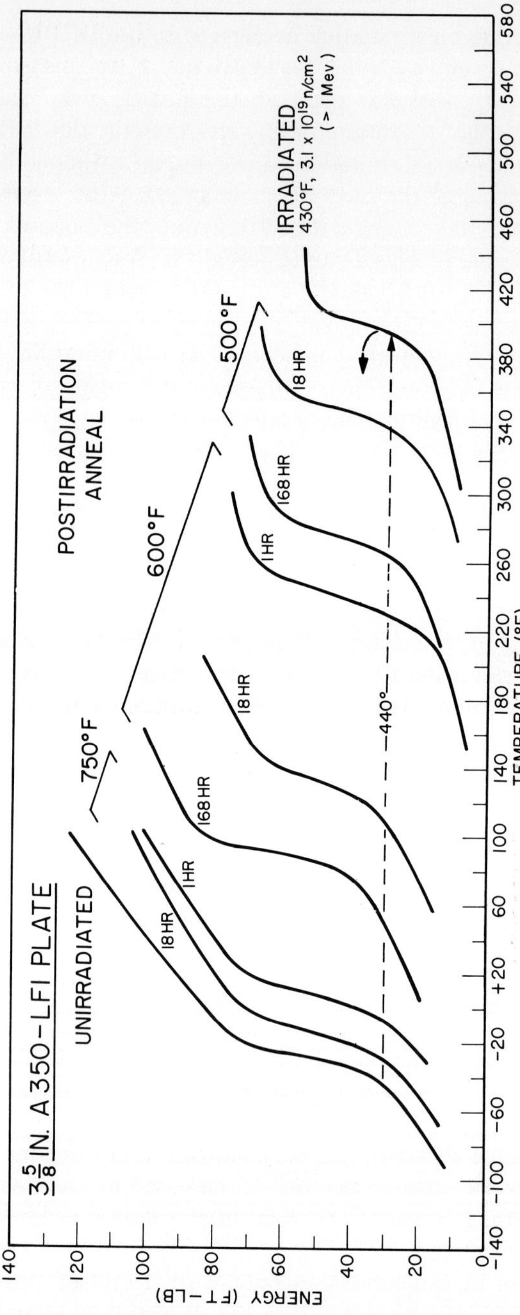

FIG. 26—Charpy V-notch ductility characteristics of irradiated A350-LF1 steel. Results of postirradiation annealing showning effects of various heat treatment (time-temperature) combinations (Ref 9).

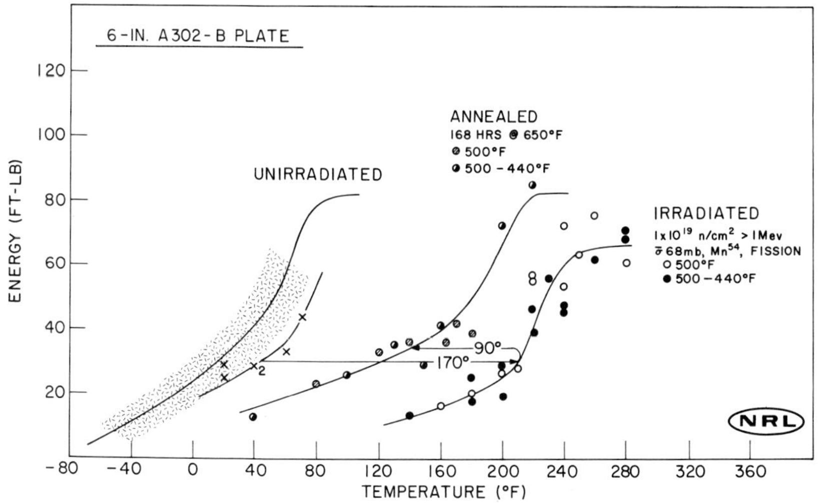

FIG. 27—*Charpy V-notch ductility behavior of ASTM A302-B steel before and after irradiation at a constant 500 F and under cyclic temperature conditions ranging from 500 to 440 F. Responses of both steel conditions to 650 F postirradiation annealing for 168 h also are shown (the* X *data points depict the unirradiated behavior of the particular section of A302-B steel used for this experiment)* (*Ref* 20).

One example is provided relative to the operation of the Yankee atomic power reactor [*20*]. At the end of each core life the reactor inlet coolant temperature was decreased gradually from about 500 to 440 F. The disadvantage of this tailing-off in temperature could be serious but is softened by the fact that at the same time the power level also is being reduced. Nevertheless, it is a point to be considered if a specific vessel is projected to receive a significant neutron exposure in its anticipated useful life. An experiment to test the effects of this type of cycle involved irradiation at 500 F and gradual reduction of temperature to simulate the Yankee cycle. The results of this experiment are shown in Fig. 27 [*20*]. No specific effect of the lowering of the irradiation to 440 F was observed. Subsequent annealing at 650 F produced about 50 percent recovery of the Charpy V-notch transition temperature with full recovery of the upper shelf energy.

Irradiation of the A302–B in the BRPR at about 585 F provides a different picture however. In this case annealing at 800 F produced 70 percent recovery in the A302–B. The variable response of the several higher strength steels to annealing at 700 and at 800 F after high fluence irradiation at about 585 F in the BRPR is quite clear (Table 2) [*19*]. These data suggest the need for annealing at about 800 F to correct significantly the embrittlement incurred at about 600 F. The extremes of irradiation embrittlement response for 585 F exposure are shown also in Table 2. Limited data on annealing of A533–B steel shows similar behavior to that of A302–B.

TABLE 2—*Summary of notch ductility characteristics of structural steels irradiated in the Big Rock Point reactor and postirradiation annealed* (*Ref* 19).

Steel	Neutron Fluence, n/cm² < 1 MeV[a]	Charpy V-notch 30-ft·lb Transition Temperature, deg F			Full Shear Energy, ft·lb		Annealed, 168 h		
		Unirradiated	Irradiated	ΔT	Unirradiated	Irradiated	Recovery	At Temperature	Percent
A302–B	7.1×10^{19}	+30	195	165	78	58	115 F	800 F	70
A543	$5.9 \times$	−130	−25	105	74	54	40 F	800 F	38
5Ni–Cr–Mo–V(3 in.)	$5.2 \times$	−165	160	325	75	51	50 F	700 F	15
5Ni–Cr–Mo–V(4 in.)	$5.5 \times$	−90	220	310	57	32	90 F	800 F	29
7–½Ni–Cr–Mo	$4.7 \times$	−215	−120	95	62	49	~70 F	800 F	74
A517–E	$6.6 \times$	−45	100	145	114	85	80 F	800 F	55
12Ni–5Cr–3Mo[b]	$7.4 \times$	−185	−55	130	64	63	~20 F	700 F	15
12Ni–5Cr–3Mo[c]	$7.3 \times$	−200	5	205	84	48	45 F	700 F	22

[a] Fission $\bar{\sigma}$ = 68 millibarns (mb), ^{54}Fe (*n*, *p*) ^{54}Mn; reactor power level, 217.3 MW.
[b] 179.1 ksi.
[c] 162.6 ksi.

Applied Stress

Since the vessel is under an applied stress while undergoing irradiation, two groups have studied the question of possible effects of stress and radiation as opposed to unstressed irradiated steels.

Reynolds [*21*] irradiated A302–B steel at a maximum stress equal to 20 percent of the yield stress and observed a slight effect of stress. Investigators at NRL [*22*] irradiated two steels, a 3⅝-in. A350-LF1 forging steel (at 430 F) and a 10½-in. A302–B plate steel (at 550 F). In both cases specimens were prestressed to 80 percent of yield and the strain checked after irradiation to assure no relaxation. Also, specimens were machined in the hot cells to assure no local stress concentration during irradiation. The results for both steels indicated no effect of stress. (See Ref *22* for detailed results.)

Hydrogen Embrittlement

The study of the combined effects of hydrogen and radiation toward the embrittlement of steels has been limited. Available results fortunately indicate no reason for concern, but the advancement of reactor materials technology suggests the need for alertness to this potentiality.

Harries and Broomfield [*23*] have studied the hydrogen embrittlement problem with respect to the service conditions of water reactor vessels and have concluded that corrosion reactions at the steel-water interface are the only significant sources of hydrogen. Further, they conclude that only by using the most pessimistic hydrogen production rate from corrosion with no vessel wall stainless cladding could serious effects of hydrogen result. Their studies also suggest that superimposed applied stress and irradiation would have, at worst, very minor effects upon the movement and effects of hydrogen in steel. With corrosion preventing cladding they believe no concern for hydrogen embrittlement need be considered at least for the water cooled reactors.

Rossin et al [*24*] investigated the combined effects of irradiation and hydrogen on the embrittlement of A212–B and 4340 steels. The results suggest that under the irradiation conditions which might be projected for a water reactor system, the A212–B is not affected seriously by even a very high content of hydrogen after irradiation. These studies, unfortunately, did not include the higher strength steels now being applied in reactor vessel service, but rather, compared the lower strength A212–B with the much higher strength 4340 steel.

A study of hydrogen embrittlement of irradiated steels is being carried out by Beeston and Brinkman [*25*] on A302–B, A542 (Cr–Mo), and A543 (Ni–Cr–Mo) steels. Their study seeks to learn the relative susceptibility of these three steels to radiation plus hydrogen embrittlement. Delayed failure characteristics are measured using notched tension speci-

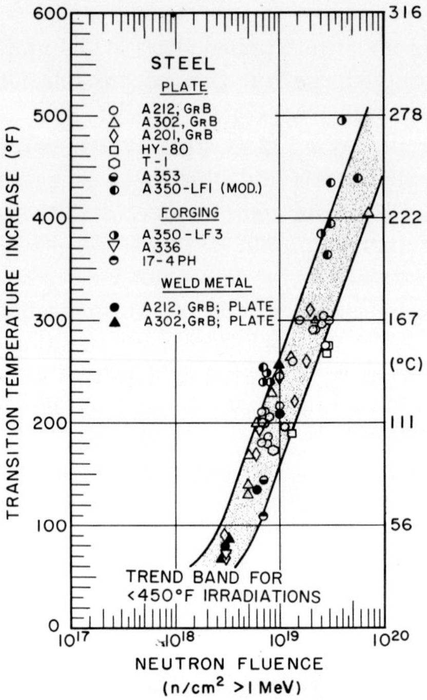

FIG. 28—*Increase in the transition temperatures of steels resulting from irradiation at temperatures below 450 F (Ref 12).*

mens which have been irradiated to various fluences then cathodically charged with various levels of hydrogen. Data on regular tensile bars are available for the A543 steel. With increasing hydrogen content, a decrease in nonuniform elongation prior to failure occurred in specimens irradiated to a fluence of approximately 2×10^{20} n/cm^2. At high hydrogen contents the ultimate stress was found to coincide with the fracture stress which was reduced below that of the ultimate strength of the irradiated but uncharged steel. The reduction-in-area values for the irradiated A542 and A543 specimens decreased to near zero with increasing hydrogen content (3 to 4 ppm).

Unfortunately, in both the work of Beeston and Brinkman and of Rossin et al, the hydrogen was injected after irradiation thereby leading to some doubt about the extent of true simulation of a vessel wall in which a mechanism for the progressive production of hydrogen may be available. Further, the rate of charging was very high compared with the highest which might be expected in service. Nevertheless, even with the severe conditions, the conclusions reached thus far in both studies suggest that delayed failure as a result of hydrogen production in pressure vessel steels cannot be projected as a significant problem with current steels under normal service conditions. With the application of higher strength steels under more severe service conditions, however, this problem may require closer attention.

Materials Factors Affecting Steel Embrittlement

Variable Embrittlement Response Among Steels

The large variations in embrittlement response which have been observed offer hope for avoiding the problem of embrittlement in the future. In the meantime, however, the extreme embrittlement (worst case) defined by the trends of Fig. 21 suggest the importance of knowing much more about the individual reactor vessel (the steel composition, the microstructure, and metallurgical condition) so that a limiting case is not the extreme of available data.

The extent of variations for a number of steels irradiated by NRL at <450 F is shown in Fig. 28 [*12*]. Adding the more recently produced steels in which better control of steelmaking practice was exercised broadens the trend to the right (Fig. 29) [*11*]. The band of data in Fig. 29 is essentially all of the 550 F or less irradiation embrittlement data gathered through 1967. Careful examination of this scatter pattern suggests several

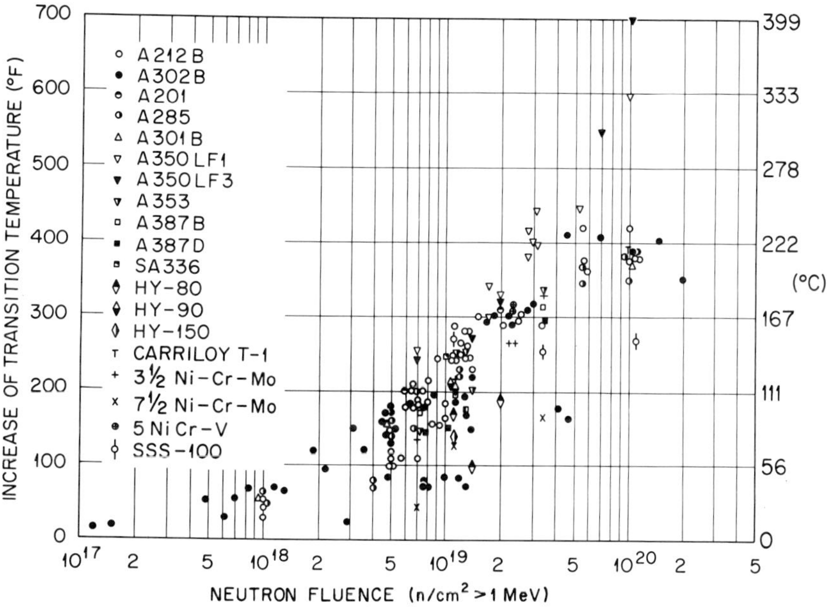

FIG. 29—*Increase in Charpy V-notch transition temperature of many steels irradiated at <500 F (Ref* 11).

facets: (1) the left side boundary (greatest embrittlement) is remarkably uniform considering the number of steels and investigators involved; (2) this boundary establishes a guide for the most conservative projection of anticipated embrittlement; (a projection which can be refined by irradiation of a specific steel at its service temperature); (3) the A350 steels (used in only one operating reactor vessel) make up the most sensitive boundary data at fluences of 1×10^{19} n/cm² and higher; (4) a great deal of scatter is evident, especially for the A302–B steel which in recent years has been improved by a vacuum pouring process, (note especially the scatter for A302–B at about 1×10^{19} with ranges of ΔT from about 70 F to well over 200 F); and (5) a general tendency for the higher strength steels to show smaller transition temperature increases. These data are useful for presenting an overall picture of steel embrittlement, but much selectivity is necessary for its practical application. However, the fact that such wide variations occur suggests that control of radiation embrittlement is possible, and recent data on special steel heats confirm this [*26, 27*].

Marked differences between individual steels are important to projecting the levels of embrittlement which may be expected in operating reactor vessels. Such individual differences are demonstrated in Fig. 30 [*28*] for

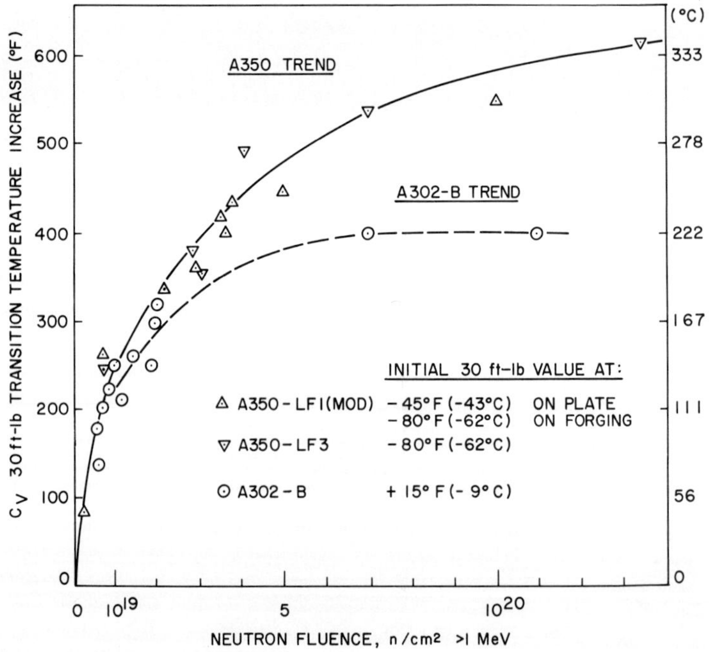

FIG. 30—*Transition temperature increases resulting from irradiation of A350 and A302-B steels to a range of fluences at temperatures below 450 F* (*Ref* 28).

TABLE 3—*Charpy V-notch ductility of four steels irradiated at 490 F to a neutron fluence of 1.4 × 10¹⁹ n/cm² >1 MeV (Ref 28).*

Material	30-ft·lb Transition Temperature, deg F		
	Initial	Irradiated	T
A350-LF3	−80	190	270
A212-B	+5	235	230
A302-B	+30	230	200
Hy-80 (Ni- Cr-Mo)	−130	−30	100

A350 and A302–B steels irradiated over a wide range of fluences. The comparable level of ΔT for four steels at the same irradiation condition are shown in Table 3 [*28*]. A similar comparison of another group of four steels after irradiation at a higher fluence and temperature is shown in Table 4 [*28*]. Highly significant variations in embrittlement sensitivity are evident in each of these comparisons, thereby suggesting the importance of knowing the sensitivity level of a particular reactor vessel steel.

A significant variable embrittlement response of steels has been noted between heats of the same steel. Out of these observations has grown a two-pronged effort to learn more about radiation sensitivity under conditions representative of anticipated service: (1) selecting commercial (or near commercial) steels having low sensitivity to radiation plus good properties for potential pressure vessel service and (2) developing steels of low (or nil) radiation sensitivity. Some results in these areas are reported.

Heat-to-Heat Variations

In one experiment [*4*], specimens of several heavy thickness heats of A302–B were irradiated simultaneously at 550 F and results compared with heats from accelerated irradiation positions in the Yankee–Rowe

TABLE 4—*Charpy V-notch ductility characteristics of four steels irradiated simultaneously at ≈550 F[a] to 3.8 × 10¹⁹ n/cm² >1 MeV (Ref 28).*

Material	30-ft·lb Transition Temperature, deg F		
	Initial	Irradiated	ΔT
A212-B	5	190	185
A302-B	30	190	160
3½ Ni-Cr-Mo	−130	−40	90
7½ Ni-Cr-Mo	−215	−200	15

[a] The average specimen temperature reached 560 F for a significant portion of the irradiation period.

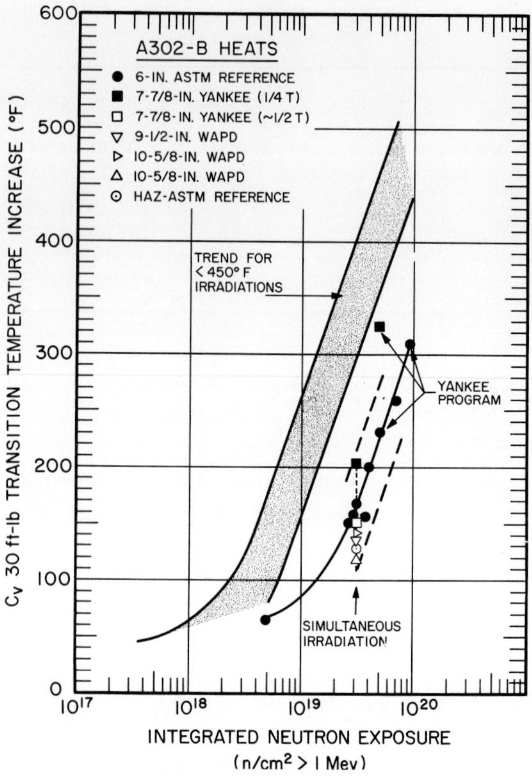

FIG. 31—*Charpy V-notch transition temperature increase for several heats of A302-B steel resulting from irradiation at 550 F in the low intensity test reactor. Six of the eight points on the vertical dashed line reference a single irradiation experiment in which specimens of five heats of A302-B were exposed simultaneously* (*Ref* 4).

Atomic Power Plant surveillance program. These data are shown in Fig. 31 [*4*]. It became apparent from this experiment that the more recently produced heavy plate A302–B steels are less sensitive to radiation embrittlement than the earlier steels, represented by the Yankee steel. The specimens for this experiment were taken from plates actually used in the vessels of nuclear plants constructed by the Westinghouse Electric Company. Comparisons between several heats of A533–B and C do not present such a uniform and desirable conclusion, however [*2*]. Data for several heats of A533 steel irradiated at 550 F produced a mixed pattern of embrittlement both greater and lesser than the reference A302–B steel. Thus, while the trend of the more recently produced A302–B heats was toward significantly reduced embrittlement sensitivity, several heats of A533 show no such consistently desirable pattern. (See discussion of plate versus weld later for more details of A533 data.)

Microstructure as a Variable in Radiation Embrittlement Sensitivity

A vivid comparison of heat-to-heat variations in embrittlement was presented by Carpenter et al [*13*]. In this study results between various heats of A302–B steel showed distinct patterns of response which were termed "sensitive" and "insensitive" (Fig. 32). The authors could not define the specific reasons for the difference but observed that the sensitive heats had "large-grain ferrite-bainite structures exhibiting some banding" while the insensitive heats had "fine-grained structures principally of ferrite and tempered lower bainite." This investigation could not define

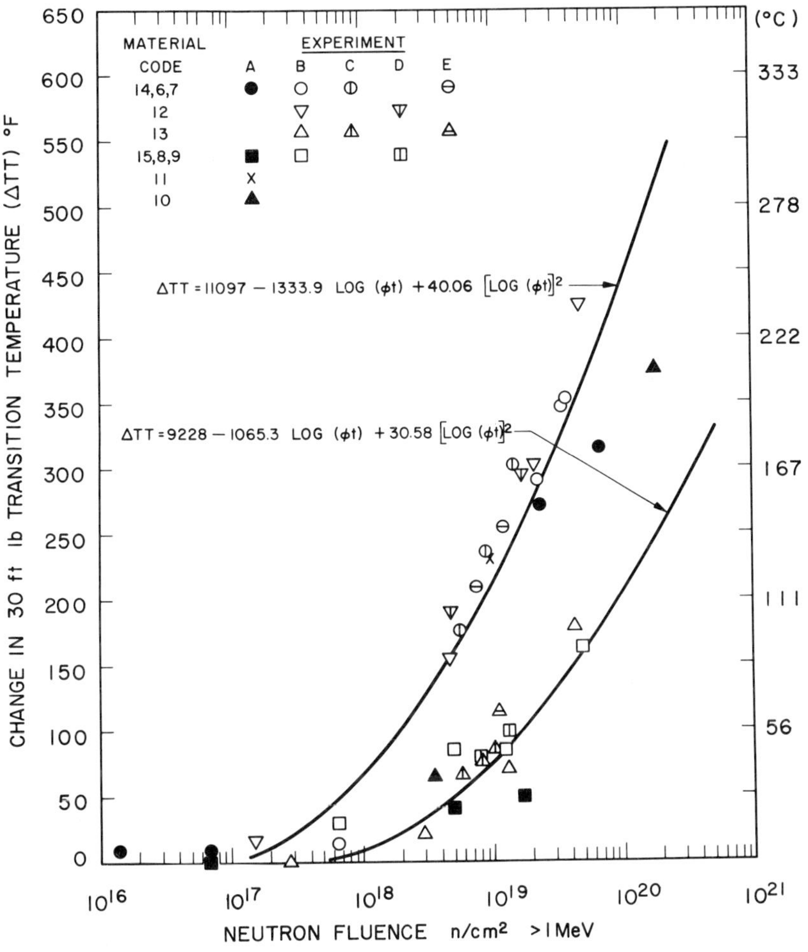

FIG. 32—*Transition temperature increases for various heats of A302-B steel at temperatures from 450 to 520 F. Embrittlement "sensitive" and "insentive" behavior trends are demonstrated by the upper and lower curves, respectively* (*Ref* 13).

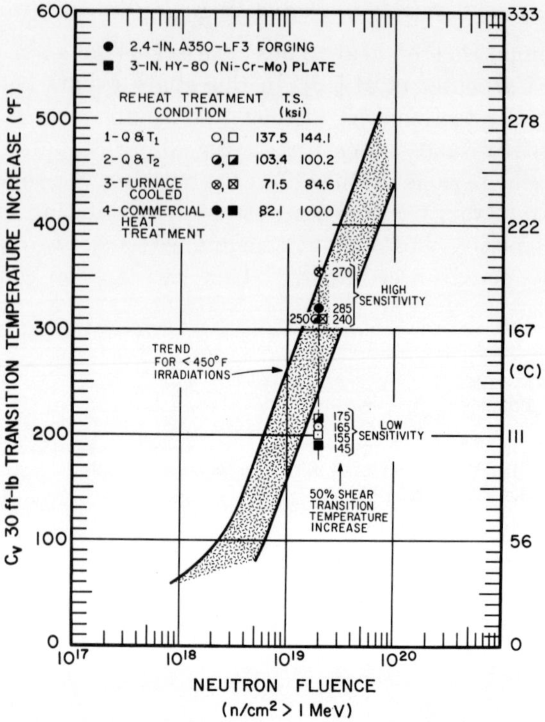

FIG. 33—*Comparison of transition temperature increases of irradiated HY-80 (Ni-Cr-Mo) and A350-LF3 steels in various heat treatment conditions. Data points depicting high radiation embrittlement sensitivity all correspond to tempered ferrite microstructure, while those denoting low sensitivity each correspond to tempered martensite.*

the influence of specific compositional variables relative to the degree of sensitivity; however, with more experience, it can be expected that critical compositional differences may be identified. (See section later on "Developing Steels for Low Radiation Embrittlement Sensitivity.") Some variation due to irradiation temperature is inherent in the results shown in Fig. 32 but do not detract significantly from the basic separation of "sensitive" and "insensitive."

Experimental investigations to isolate and assess metallurgical factors responsible for variable radiation embrittlement sensitivity among reactor steels have included a study of microstructural effects [*26*]. The study proceeded with two steels, HY-80 (Ni-Cr-Mo) and A350-LF3, which in their normal "commercial" heat treatment conditions, respectively, set the bounds of low- and high-embrittlement sensitivities as found by NRL in evaluations of several low- and medium-strength structural steels. The steels were individually reheat treated to four different strength levels from

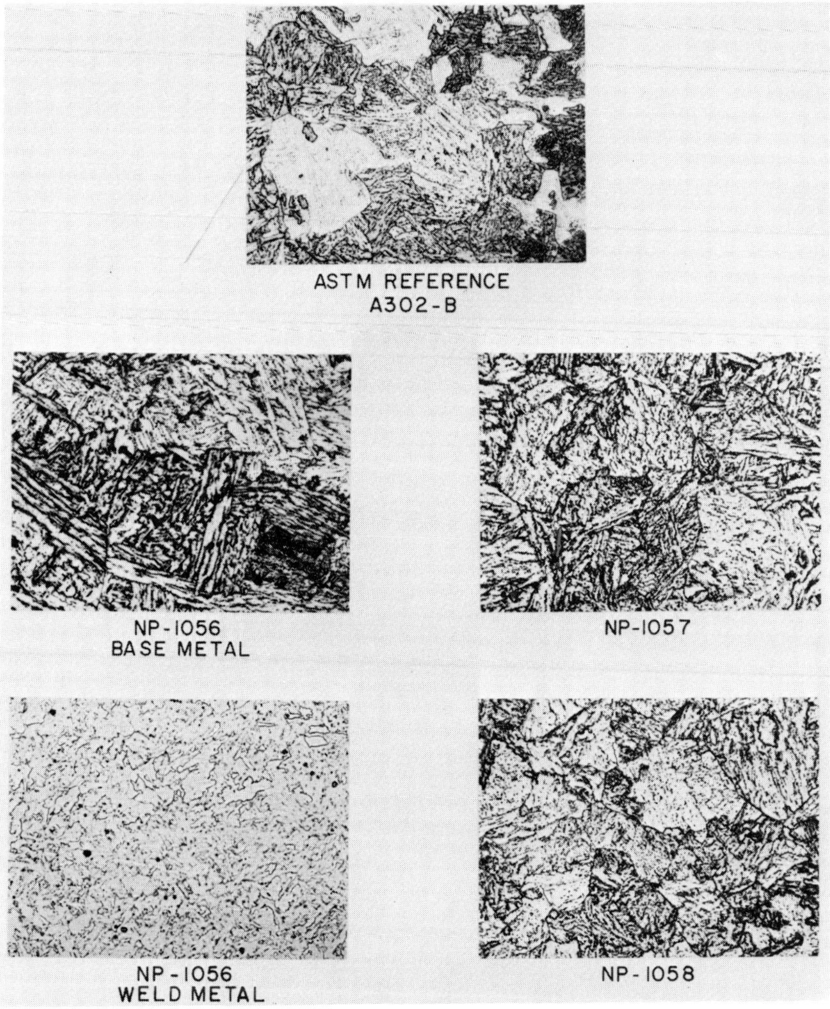

FIG. 34—*Microstructures of the ASTM A302-B reference plate, the three LaCrosse reactor A302-B plates, and of the weld metal from the LaCrosse reactor fabrication test plate. The microstructure of all four plates are tempered upper bainite; the weld metal structure is ferrite with spheroidized carbide particles. Nital-picral etchant* (× *500*) (*Ref* 29).

70 to 145 ksi wherein the general microstructure of both steels ranged from tempered ferrite to tempered martensite.

A comparison of the Charpy V-notch transition temperature responses of both steels to <250 F irradiation is shown in Fig. 33 for each of the four heat treatment conditions. The data not only revealed an appreciable change in the radiation embrittlement sensitivities of both steels on varying their heat treatment condition, but also a complete reversal in the order of their relative irradiation response levels. More important, the transition from high to low sensitivity in each case was found to correspond to the change in general microstructural appearance from tempered ferrite (high sensitivity) to tempered martensite (low sensitivity). Thus, microstructure is apparently one key metallurgical variable influencing radiation embrittlement sensitivity of low- and medium-strength steels as characterized by the steels used in this investigation. In the quenched and tempered condition, however, significant microstructural variations in Type A533 steels are not expected.

Variations Between Plate, Weld, and Weld Heat Affected Zone

The variations in radiation sensitivity within a welded vessel of a given type of steel are of more importance than the academic question of variations between steels. This statement is amplified by observations to date on A302–B and A533 steels which indicate that welds are more sensitive than plate and the plate form is somewhat more sensitive than HAZ.

Early studies of plate and welds showed no significant embrittlement difference (Fig. 28). However, these comparisons were too limited in materials and irradiation temperatures to show the differences which have been observed more recently.

Examination of the BRPR vessel surveillance specimens showed big differences in irradiation response between plate and weld, the weld being more sensitive, though even the weld changes were small. Detailed results are given in Chapter 4.

Before the LaCrosse (Wisconsin) boiling water reactor (LACBWR) reached operational status, sections of three heats of the vessel steel were provided for irradiation study [*29*]. The compositions of the steels involved plus that of the A302–B reference steel are shown in Table 5. The microstructures of the five steels involved are shown in Fig. 34 [*29*]. Representative specimens of the three LaCrosse plates along with the weld and the HAZ of one LaCrosse plate were irradiated to two different fluences, 1.0 and 2.1×10^{19}. The reference plate was irradiated to 2.1×10^{19}. The results shown in Table 6 indicate greater sensitivity of the weld with superiority in the HAZ. However, all of the LaCrosse steels were superior to the reference steel. These data call attention to another point, however, the relatively low shelf values of the weld and of the plates in transverse orientation. The weld, a test weldment fabricated according to LACBWR shop practice, showed considerably lower energy absorption values than the plate but suffered little reduction as a result of irradiation at 550 F (Fig. 35) to 2.1×10^{19} n/cm^2. Nevertheless, this must be considered the weakest aspect of this plate when welded.

The general superiority of the BRPR and LACBWR plates are traced to careful production with control of residual element content. An important factor in radiation sensitivity has been shown to be the copper and

TABLE 5—*Chemical composition and heat treatment of 4-in.-thick A302–B LaCrosse reactor pressure vessel plate and test weldment and of 6-in.-thick A302–B ASTM reference plate* (*Ref* 29).

A. CHEMICAL COMPOSITION

Heat Number	C	Mn	Si	P	S	Ni	Cr	Mo	V	Al	Cu
NP–1056											
Base	0.22	1.35	0.22	0.007	0.018	0.18	0.06	0.52	0.02	0.01	0.11
Weld	0.10	1.39	0.43	0.016	0.006	0.12	0.03	0.55	0.02	0.01	0.18
NP–1057	0.21	1.31	0.19	0.006	0.018	0.10	0.05	0.52	0.02	0.01	0.11
NP–1058	0.20	1.26	0.25	0.008	0.006	0.16	0.07	0.55	0.02	0.01	0.16
ASTM reference	0.24	1.34	0.23	0.011	0.023	0.18	0.11	0.51	...	0.04	0.20

B. HEAT TREATMENT

Plates from Heats NP–1056, 1057, and 1058: Solution annealed at 1950 to 2050 F; water quenched; austenitized at 1725 to 1775 F, held 1 h/in. of thickness minimum, water spray quenched to 500 F; tempered at 1200 to 1250 F, 2 h air cooled.

Due to an inadvertent subsequent overheat, plate from Heat NP–1058 was reheat treated as follows: Heated to 1750 F, held 4 h at temperature, quenched in well agitated oil to 200 F; tempered at 1225 F for 6 h, air cooled.

Weldment: Preheat temperature, 300 to 500 F. Postwelding stress relief annealing temperature, 1100 to 1175 F.

Stress relief annealing: All plates were subjected to at least two stress relief cycles consisting of: heating to the temperature range of 1100 to 1175 F and holding for 1 h/in. of plate thickness, furnace cooling to 500 F, and air cooling.

TABLE 6—*Summary of Charpy V-notch ductility evaluations on LaCrosse pressure vessel plates and weldments* (*Ref* 29).

Plate Code	Thickness Location	Orientation in Plate	Neutron Fluence, n/cm²[a] < 1 MeV	Charpy V-notch 30-ft·lb Transition Temperature, deg F			Energy Absorption at Full Shear, ft·lb		
				Unirradiated	Irradiated	ΔT	Unirradiated	Irradiated	Δft·lb
				HEAT NP 1056					
LC1.......	¼ to ¾	longitudinal	2.1 × 10^{19}	35	120	85	91[b]	81	
LC1.......	weld	...[c]	2.1	10	120	110	60[b]	60	
LC1.......	HAZ	fusion line	2.1	−65	0	65	125	~80	~45
				HEAT NP 1057					
LC2.......	¼ to ¾	longitudinal	2.1	−35	~−25	<20	143	129	14
LC2.......	¼ to ¾	transverse	1.0	−5	25	30	72	66	6
LC3.......	¾	longitudinal	1.0	−45	−45	0	149	141	8
				HEAT NP 1058					
LC4.......	¼ to ¾	longitudinal	2.1	−25	40	65	150	120	30
LC4.......	¼ to ¾	transverse	1.0	25	85	60	76	65	11
LC5.......	¾	longitudinal	1.0	−35	20	55	144	144	0

[a] $\bar{\sigma}$ 68 millibarns (mb), Mn^{54}, fission spectrum.
[b] >90 percent shear.
[c] Notch parallel to plate thickness; specimen perpendicular to weld length.

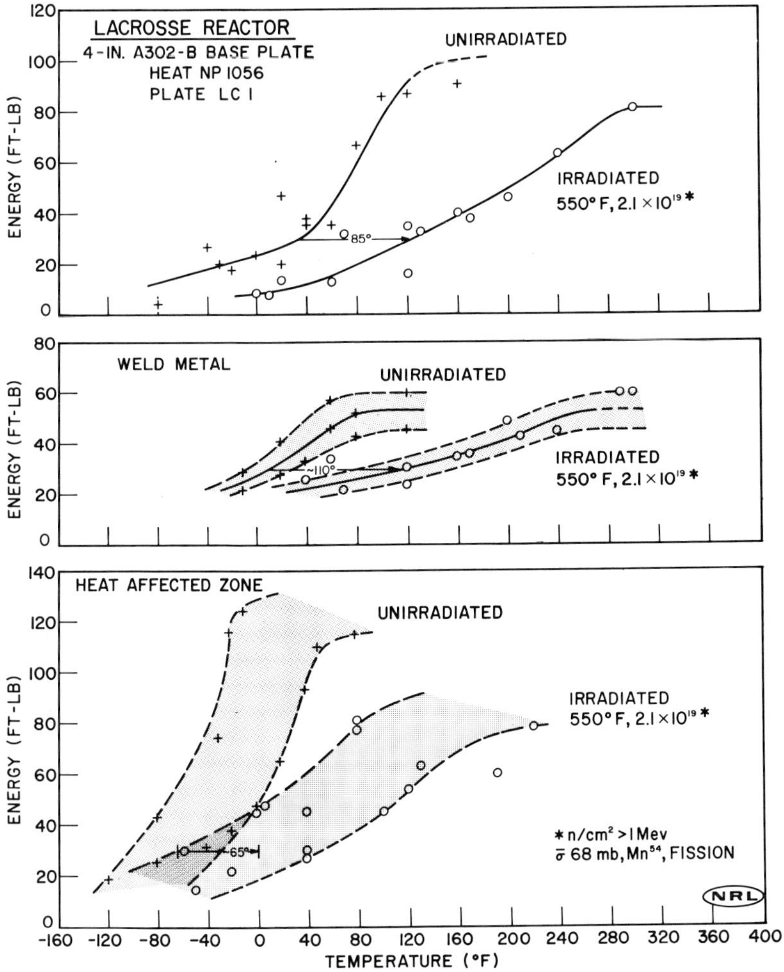

FIG. 35—*Charpy V-notch ductility characteristics of 4-in. LaCrosse reactor vessel fabrication test weld including weld metal, base plate (Heat NP-1056), and weld heat affected zone (HAZ) properties, before and after irradiation at 550 F to 2.1 × 10^{19} n/cm² > 1 MeV (Ref* 29).

phosphorus content [30]. Looking at Table 5 it is possible to trace a pattern of radiation sensitivity with copper content, the reference steel being most sensitive and highest in copper and the plates of Heats NP 1056 and 1057 being least sensitive and lowest in copper. This factor of composition is discussed further later.

Initial Results on A533 Steel Plate and Welds

Early comparisons of irradiation effects data for various A533 plates and welds show a mixed pattern of results which destroy the favorable

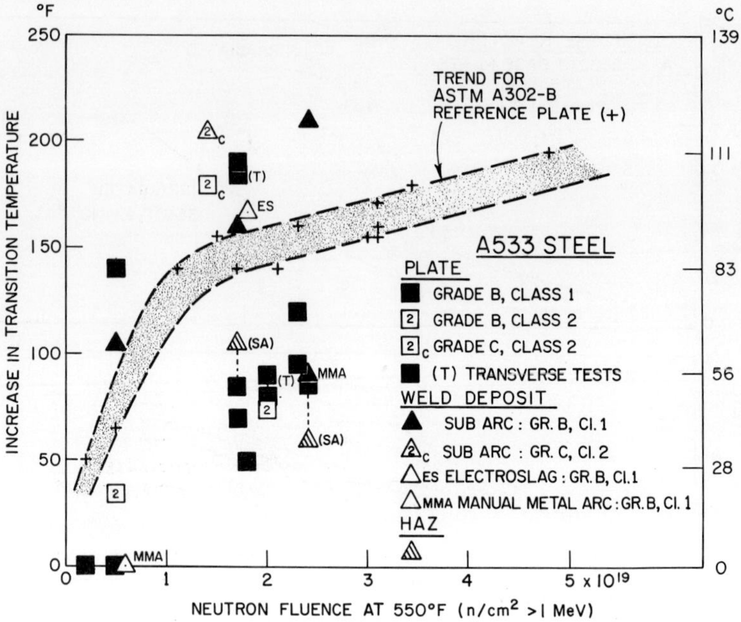

FIG. 36—*Comparison of transition temperature increases for various A533 steel plates and welds in relation to trend established for irradiation of reference A302-B steel. Exposure temperature was 550 F for all data (based on data from Ref* 27).

impression produced by the BRPR and LACBWR results as to the radiation sensitivity of newer steels. Embrittlement data [*2*] for a number of A533 plates (including both Grades B and C and Strength Classes 1 and 2) and welds irradiated to various fluences at 550 F are shown in Fig. 36. It can be readily seen that sensitive and insensitive conditions of both plate and weld were observed. Several plates from the manual metal arc weld and the weld heat affected zone specimens (limited data) showed less embrittlement than the A302–B reference steel. The electroslag weld though much more sensitive than the related plate (Fig. 37) [*2*] proved no more sensitive than certain submerged arc welds.

The several plates and welds examined (Fig. 36) represent recent and typical vessel construction. The promise for lower sensitivity which was so apparent in the trends of Figs. 31 and 32 has not been borne out by these early A533 data. It should be noted, however, that the most sensitive A533 is greatly superior to the imposed (conservative) industry design basis (Fig. 19). Vessel surveillance receives renewed importance from such a divergent trend of results among the commercial A302 and A533 steels.

The most vivid variation between plate and weld sensitivity observed to date is for the A543 steel. A number of plates of HY–80 or A543 steel have been irradiated. In general, the plate form has been shown to be

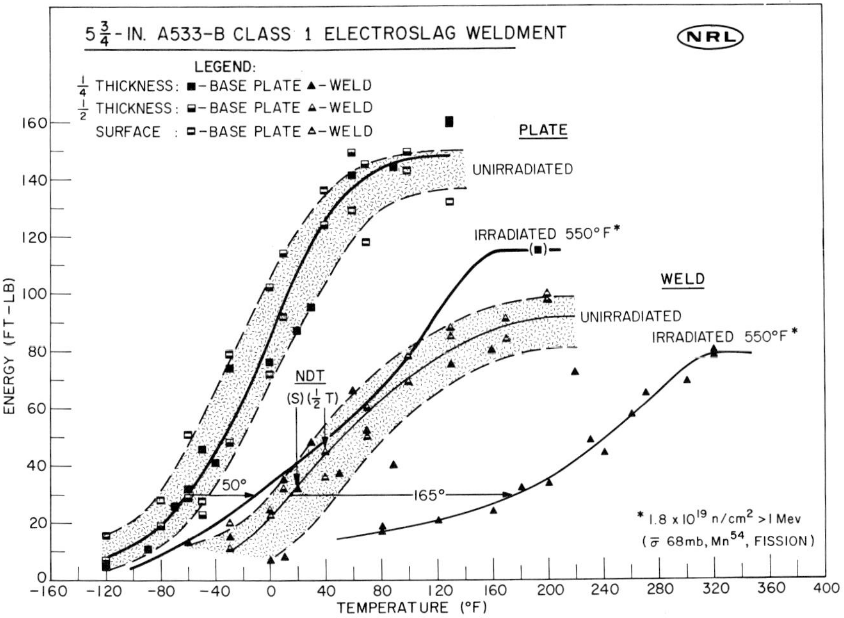

FIG. 37—*Charpy V-notch ductility results for 5¾-in. A533-B Class 1 electroslag weld and related plates. Notch ductility properties of the parent plate are markedly superior to those of the weld metal both before and after irradiation* (*Ref* 2).

superior to A212 and A302 compositions, as indicated previously in Table 3. Because the plate form was so good it was decided that weld metal should be also examined. The results were shocking; the weld proved to be the most sensitive of all the steels (weld or plate) examined to date at NRL. A comparison of A543 (HY–80) plate and weld embrittlement is shown in Fig. 38 [*31*]. The very wide differences in sensitivity led to an examination of possible causes of the differences. One obvious difference is that of copper content as was the case with the LACBWR steels, and, again, there was a direct parallel between high embrittlement and high-copper content versus low embrittlement and low-copper content in the base plate. Additional investigations [*2*] are aimed at improvements in the weld metals for reactor vessel steels. The large variations in sensitivity described above have built-in suggestions for control, a subject which is now being pursued in depth.

Developing Steels for Low Radiation Embrittlement Sensitivity

Several investigators have studied the effect of grain size on radiation sensitivity. Nichols and Harries [*18*] described grain size effects for five steels. Their results are summarized in Table 7 which also lists the compositions. For Steel 2, for example, the heat treatment was varied to con-

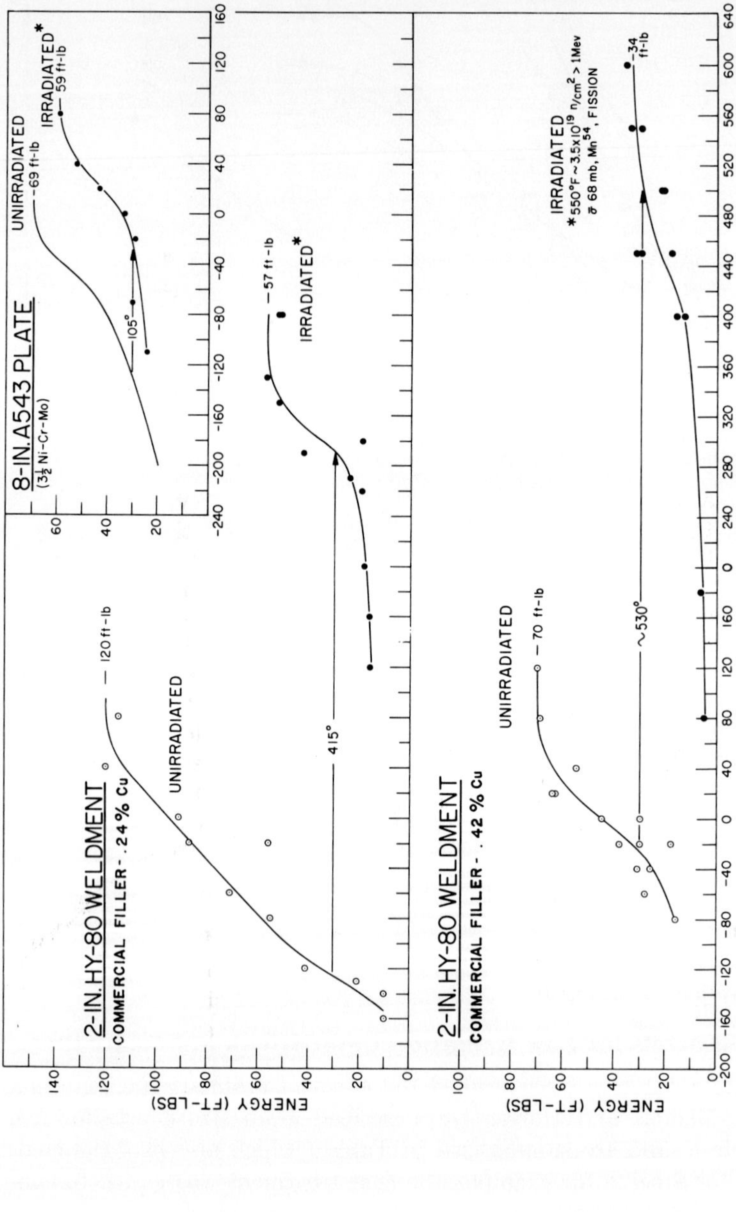

FIG. 38—*Charpy V-notch ductility of 2-in. HY-80 weldments (submerged arc commercial filler) and 8-in. A543 plate before and after irradiation at 550 F (Ref 31).*

TABLE 7—*Effect of grain size upon radiation sensitivity of three steels* (*Ref* 18).

Steel[a]	Heat Treatment	Ferrite Grain Size, grains/mm²	Increase in Transition Temperature, 35 ft·lb	
			deg C	deg F
Aluminium grain size controlled mild steel[a (1)]	1000 C (1832 F) for 1 h; transferred to 900 C (1652 F), cooled 40 C (104 F)/h to 650 C (1202 F), held 16 h, furnace cooled	523	50	122
	930 C (1706 F) for ½ h; cooled at 40 C (104 F)/h to 650 C (1202 F), held 16 h, furnace cooled	6 890	5	41
Silicon-killed mild steel[a (2)]	1100 C (2012 F) for 8.5 h; transferred to 930 C (1706 F); cooled 12 C (54 F)/h to 650 C (1202 F), held 16 h, furnace cooled	76	65	149
	930 C (1706 F) for ½ h; cooled at 40 C (104 F)/h to 650 C (1202 F), held 16 h, furnace cooled	842	45	113
	930 C (1706 F) for ½ h; air cooled to room temperature, 650 C (1202 F) for 16 h, furnace cooled	4 130	30	86
Experimental nickel-molybdenum-vanadium steel[a (3)]...	930 C (1706 F) for ½ h; cooled at 40 C (104 F)/h to 650 C (1202 F), held 16 h, furnace cooled	5 060	40	104
	930 C (1706 F) for ½ h; air cooled to room temperature, 650 C (1202 F) for 16 h, furnace cooled	36 500	10	50

[a] Analysis of steels:

(1) 0.13% C; 0.12% Si; 1.19% Mn; 0.08% Ni; <0.05% Cr; 0.06% Mo; <0.03% V; 0.19% Cu; 0.012% Al; 0.004% O; 0.005% N.
(2) 0.12% C; 0.09% Si; 1.26% Mn; 0.07% Ni; <0.05% Cr; <0.05% Mo; <0.03% V; 0.19% Cu; 0.005% Al; 0.013% O; 0.005% N.
(3) 0.11% C; 0.12% Si; 1.05% Mn; 1.04% Ni; 0.07% Cr; 0.24% Mo; 0.17% V; 0.08% Cu; 0.007% Al; 0.003% O; 0.006% N.

trol grain size and obviously aids in reducing the sensitivity to radiation embrittlement. However, composition variations may also have important effects. For example, Steel 3 contains low copper compared to the other two steels but contains more vanadium. Certain levels of both copper and vanadium have been shown to contribute to steel sensitivity [*27*]. Gordon

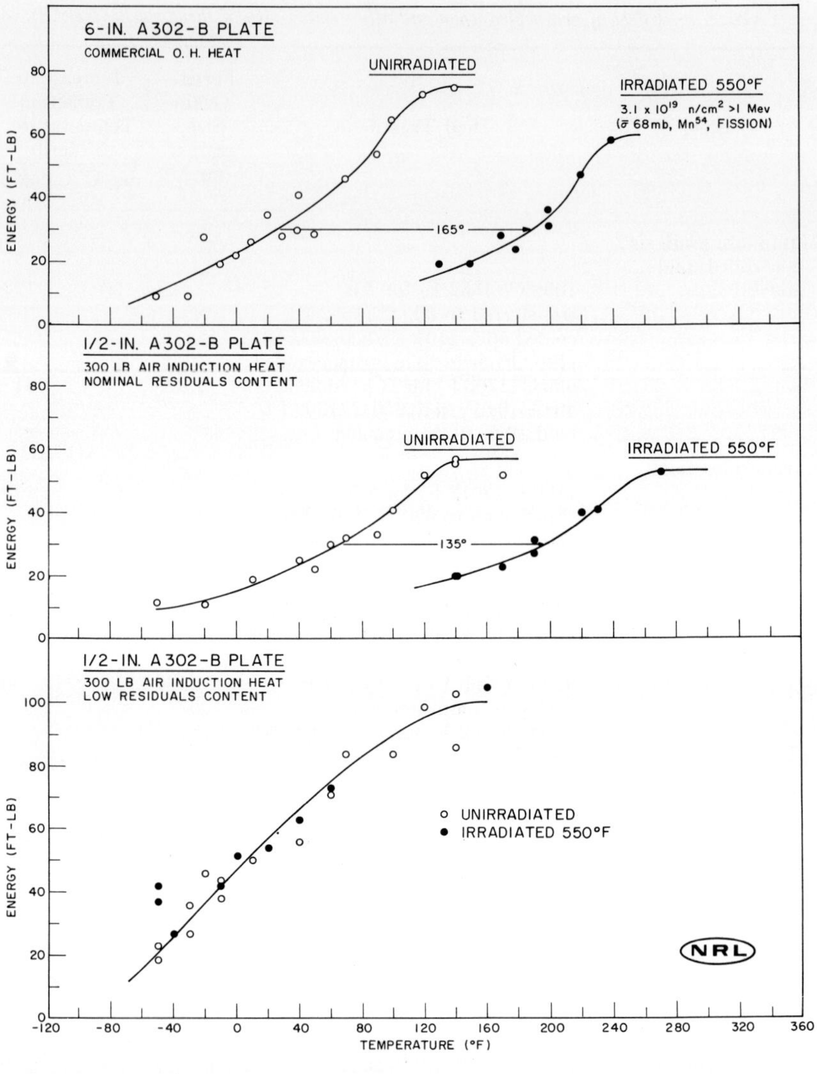

FIG. 39—*Comparison of radiation embrittlement sensitivities of one large commercial heat and two air induction heats of A302-B steel with nominal and low residual element contents following 550 F irradiation* (*Ref* 27).

and Klepfer [*32*] also have described grain size control as being most important to the control of radiation embrittlement.

It is clear that both composition and microstructure affect the embrittlement sensitivity of a steel, and this suggests that the controlled production of steels should permit the development of steels of low sensitivity to radiation embrittlement. Such an effort has been carried on at

NRL for several years with success. The NRL effort was stimulated by an AEC request to test the relative radiation embrittlement of uranium and nonuranium bearing A212–B steels. The surprising results [9], low initial transition and low sensitivity for the uranium steel, led to a further examination of various heats of laboratory-produced steel (Type A302–B). The A212–B steel with uranium had been produced with special care including an unusual vacuum degassing step.

The next effort was a joint effort with the U.S. Steel Applied Research Laboratory to produce A302–B steel with low residual element content, followed by a major investigation of individual compositional factors and radiation embrittlement. Figure 39 [27] indicates that effective insensitivity to irradiation at 550 F was accomplished in an air induction heat having low residual element content (that is, minimum content of elements such as phosphorus, sulfur, arsenic, copper, tin, antimony, etc.).

The next step was to determine the effects of individual residual elements through an intricate series of controlled chemistry laboratory heats, the composition and heat treatments for which are listed in Table 8. Irradiation experience and general knowledge of troublesome elements in steel making led to control of copper, vanadium (Heat Y9238); nitrogen, copper, vanadium (Y9239); and sulfur, phosphorus (Y9240). Other residual elements were controlled at minimum levels. The results of irradiation on the notch ductility of these ten steels are summarized in Table 9. Metallographic examination and aging studies also were carried out. Conclusions from these studies include:

1. Laboratory heats of A302–B steel of controlled residual element content can be produced at will to show low radiation sensitivity.

2. Such heats are consistently less sensitive to irradiation than the commercial A302–B even when "sensitizing" elements are added.

3. Higher sensitivity is produced by the addition of copper (~0.20 percent), vanadium (~0.08 percent), phosphorus (~0.22 percent plus sulfur) to the low sensitivity composition.

4. Addition of sulfur, though it changes the transition temperature characteristics of the unirradiated steel, has no apparent effect upon sensitivity to radiation.

5. Nitrogen contents between 0.008 and 0.15 percent in the A302–B laboratory compositions had no measureable effect upon sensitivity.

6. Metallographic examinations to ×1000 show structures of proeutectoid ferrite and tempered upper bainite in all laboratory heats with little distinguishable differences attributable to the additions except the identification of manganese sulfide inclusions in Heat Y9240 to which sulfur was added. (Electron microscopic examination studies will test further the possibility of microscopic differences which might aid the understanding of the influence of the several added elements upon radiation sensitivity.)

TABLE 8—*Chemical analysis and heat treatment of controlled chemistry laboratory heats of A302-B steel and of the commercial reference heat (Ref 27).*

Material Identification	Heat Treatment	Chemical Analysis, weight %[a]												
		C	Mn	P	S	Si	Mo	Cu	V	Al		N		0
										Soluble	Insoluble	Ester-halogen	Total	
Heat Y9238–Cast 1A...	A	0.23	1.40	0.003	0.004	0.28	0.52	<0.005	<0.005	0.029	0.006	...	0.005	0.0057
		0.24	1.34	0.003	0.004	0.27	0.54	0.002	0.010	0.027	0.005	0.005	0.009	...
Heat Y9238–Cast 1B (Cu added)..........	A	0.23	1.38	0.003	0.004	0.28	0.52	0.21	<0.005	0.026	0.004	...	0.006	0.0032
		0.24	1.34	0.004	0.004	0.24	0.54	0.20	0.010	0.021	0.006	0.005	0.008	...
Heat Y9238–Cast 1C (Cu and V added)....	B	0.25	1.32	0.003	0.004	0.26	0.51	0.20	0.084	0.016	0.004	...	0.006	0.0045
		0.25	1.26	0.004	0.004	0.22	0.54	0.20	0.080	0.016	0.006	0.005	0.011	...
Heat Y9239–Cast 1A (high N)............	A	0.26	1.32	0.003	0.003	0.21	0.50	<0.005	<0.005	0.029	0.005	...	0.013	0.0059
		0.26	1.24	0.004	0.005	0.21	0.54	0.002	0.010	0.027	0.005	0.012	0.017	...
Heat Y9239–Cast 1B (high N and Cu added)..............	A	0.26	1.28	0.003	0.003	0.22	0.50	0.20	<0.005	0.025	0.005	...	0.014	0.0048
		0.25	1.25	0.004	0.005	0.20	0.53	0.20	0.010	0.021	0.005	0.011	0.017	...
Heat Y9239–Cast 1C (high N, Cu and V added)..............	B	0.26	1.26	0.003	0.004	0.19	0.50	0.20	0.076	0.016	0.006	...	0.014	0.0053
		0.25	1.25	0.004	0.005	0.19	0.53	0.20	0.080	0.016	0.005	0.008	0.014	...

Heat Y9240–Cast 1A...	A	0.27	1.39	0.004	0.004	0.27	0.50	<0.005	<0.005	0.029	0.007	...	0.009	0.0051
		0.26	1.30	0.006	0.005	0.23	0.54	0.002	0.010	0.027	0.005	...	...	...
Heat Y9240–Cast 1B (S added)...........	A	0.26	1.39	0.004	0.021	0.27	0.50	<0.005	<0.005	0.030	0.004	...	0.010	0.0038
		0.24	1.33	0.006	0.021	0.23	0.54	0.002	0.010	0.027	0.006	...	.	...
Heat Y9240–Cast 1C (S and P added).....	A	0.24	1.38	0.031	0.021	0.26	0.50	<0.005	<0.005	0.022	0.004	...	0.010	0.0042
		0.23	1.30	0.021	0.022	0.21	0.53	0.008	0.010	0.026	0.005	...	.	...
A302-B reference heat (6-in. plate).........	C	0.24	1.34	0.011	0.023	0.23	0.51	0.20	<0.001	0.038	0.002	...	0.008	...
		0.23	1.35	0.015	0.022	0.22	0.52	0.22	<0.010	0.02	...	...	0.008	

NOTE—Heat Treatment A: Austenitized at 1650 F for 1 h, air cooled; tempered at 1200 F for 1 h, water quenched. Heat Treatment B: Austenitized at 1650 F for 1 h, air cooled; tempered at 1275 F for 6 h, water quenched. Heat Treatment C: Austenitized at 1650 F for 2 h, water quenched; tempered at 1200 F for 6 h, furnace cooled to below 600 F.

[a] First number = courtesy supplier; second number = NRL determination.

TABLE 9—*Charpy V-notch ductility characteristics of controlled chemistry laboratory heats of A302-B steel and of the commercial reference heat before and after irradiation at 550 F (Ref 27).*

Material Identification	Neutron Fluence, n/cm^2 > 1 MeV × 10^{19}	C_v 30-ft·lb Transition Temperature, deg F			Approximate Full Shear Energy Absorption, ft·lb	
		Unirradiated	Irradiated	ΔT	Unirradiated	Irradiated
Heat Y9238–Cast 1A	2.3	−40	−40	0	120	120
Heat Y9238–Cast 1B (Cu added)	2.3	−55	0	55	140	130
Heat Y9238–Cast 1C (Cu and V added)	2.3	60	155	95	140	135
Heat Y9239–Cast 1A (high N)	2.3	−50	−50	0	125	125
Heat Y9239–Cast 1B (high N, Cu added)	2.3	−50	30	80	135	130
Heat Y9239–Cast 1C (high N, Cu and V added)	2.3	25	90	65	130	140
Heat Y9240–Cast 1A	3.0	−20	−20	0	125	125
Heat Y9240–Cast 1B (S added)	3.0	0	0	0	75	75
Heat Y9240—Cast 1C (S and P added)	3.0	20	100	80	75	75
A302–B reference heat (6-in. plate)	3.0	15	170	155	82	67

TABLE 10—*Chemical composition and heat treatments of special A533–B steel heat* (*Ref* 19).

Specimen	Chemical Analysis, weight %															
	C	Mn	P	S	P + S	Si	Ni	Mo	Cr	Cu	V	As	Sb	Sn	Bi	As + Sb +Sn+Bi
Melt aim[a].....	0.20	1.30	0.007 max	0.007 max	0.012 max	0.24	0.55	0.55	LAP[b] 0.05 max	LAP[b] 0.08 max	LAP[b]	LAP[b] 0.01 max	LAP[b] 0.01 max	LAP[b] 0.02 max	LAP[b] 0.02 max	0.05 max
Latrobe steel hot top analysis:																
Ingot 1.....	0.18	1.38	0.006	0.006	0.012	0.20	0.56	0.53	0.06	0.12 (0.13% aim)	...	<0.01	<0.01	<0.02	<0.02	<0.05
Ingot 2.....	0.18	1.37	0.006	0.006	0.012	0.20	0.56	0.53	0.06	0.03	...	<0.01	<0.01	<0.02	<0.02	<0.05

HEAT TREATMENTS

Plate No. 1 (Class 1): Austenitized at 1675 F for 1 h/in., quenched in agitated water; reaustenitized at 1575 F for 1 h/in., quenched in agitated water; tempered at 1250 F for 1 h/in., furnace cooled; stress relief annealed at 1125 F for 20 h, furnace cooled.

Plate No. 2 (Class 2): Austenitized at 1675 F for 1 h/in., quenched in agitated water; reaustenitized twice at 1575 F for 1 h/in., quenched in agitated water; tempered at 1150 F for 1 h/in., air cooled; stress relief annealed at 1110 F for 10 h, quenched in agitated water; restress relief annealed at 1050 F for 6 h, furnace cooled.

[a] As specified by NRL and prime contractor (Lukens Steel Corp.).
[b] LAP .. low as possible.

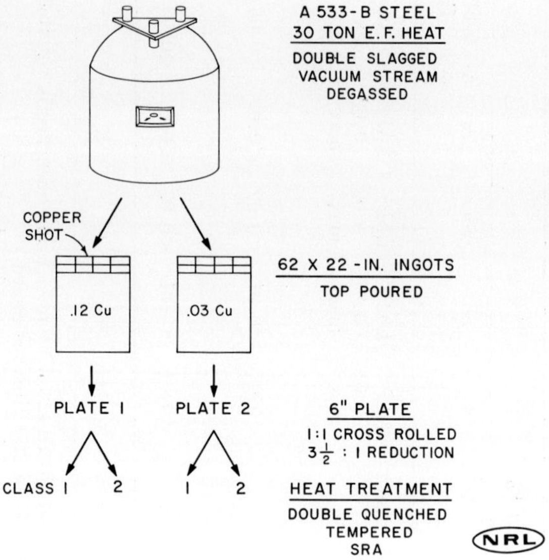

FIG. 40—*Schematic plan of processing of special commercial heat of A533-B steel for radiation sensitivity studies* (*Ref* 19).

7. These results when related to prior irradiation effects studies were convincing enough that specifications were written for a 30 ton (commercial) electric furnace heat. Specifications and heat treatment are shown in Table 10; breakdown of the heat is shown schematically in Fig. 40 [*19*]. Examination of the radiation sensitivity of this heat was begun in early 1969.

Looking to future pressure vessels, the results on the special A302–B steels were applied in an attempt to reduce the sensitivity of weld metals of the A543 steel (Fig. 38).

Welds of selected compositions were studied in cooperation with the Babcock and Wilcox Company [*27*]. The compositions of these experimental and two commercial weld deposits are shown in Table 11 with the welding parameters and heat treatment of the welds in Table 12. The results (Table 13) after irradiation were most striking in terms of reduced sensitivity for the experimental welds containing lower copper contents. For graphic contrast, the results for experimental weld 1936 and the commercial weld (0.42 Cu) are shown in Fig. 41 along with an 8-in. A543 reference plate [*31*]. It is clear that at least on an experimental basis it is possible to reduce the sensitivity basis for limiting embrittlement in future vessels, but the basis must be proven on a commercial scale next; and, this is underway for both A533 and A543 in both plate and weld form.

TABLE 11—*Chemical analysis of experimental and commercial weld deposits and of the A453 reference plate* (*Ref* 27).

Material Identification	Analysis, weight %[a]												
	C	Mn	P	S	Si	Cr	Ni	Mo	Cu	V	As	Sn	Sb
Experimental composition 1934	0.04	0.96	0.010	0.016	0.45	2.04	0.79	1.01	0.14	0.02	...	...	...
	0.04	0.87	0.008	0.010	0.40	2.08	0.77	1.01	0.06[b]	0.02	0.005	<0.007	<0.01
Experimental composition 1936	0.06	0.94	0.012	0.016	0.26	2.12	0.71	1.02	0.02	0.02	...	...	...
	0.06	0.86	0.007	0.010	0.30	2.08	0.78	1.01	0.02[b]	0.02	0.007	<0.007	<0.01
Experimental composition 1938	0.03	0.74	0.011	0.017	0.32	2.39	1.95	0.99	0.14	0.02			
	0.03	0.72	0.009	0.010	0.34	2.39	1.62	0.92	0.07[b]	0.03			
Experimental composition 1948	0.05	0.98	0.012	0.017	0.22	2.31	1.66	0.94	0.02	0.02			
	0.05	0.82	0.007	0.006	0.20	2.32	1.56	0.92	0.05[b]	0.03			
Commercial filler (0.24 Cu)	0.06	1.24	0.011	0.009	0.37	0.17	1.25	0.33	0.24	...			
	0.07	1.07	0.011	0.006	0.37	0.35	1.58	0.37	0.24	...			
Commercial filler (0.48 Cu)	0.09	1.38	0.013	0.011	0.27	0.32	2.13	0.48	0.43	...	...	...	...
	0.09	1.15	0.011	0.006	0.22	0.39	2.46	0.52	0.42	...	0.010	<0.007	<0.01
8-in. A543 plate	0.17	0.38	0.013	0.023	0.29	1.88	3.65	0.51	...	...			
	0.17	0.34	0.013	0.015	0.25	1.93	3.45	0.51	0.06	0.02			

[a] First number = courtesy supplier; second number = NRL determination.
[b] Average of three check analyses.

NOTE—The 8-in. A543 base plate was heat treated as follows: 1650 F/2 h, water quench. 1500 F/2 h, water quench. 1185 F/2 h, water quench. Welding and heat treatment conditions for all weldments are given in Table 6.

TABLE 12—*Welding Parameters and Heat Treatment of the Experimental and Commercial Weldments (Ref 27).*

Experimental Weld Compositions 1934, 1936, 1938, and 1948	
Welding process	manual metal arc
Electrode	5/32 dia, Type 15 coating
Current	165 A
Voltage	25 V
Travel	6 in./min
Preheat temperature	300 F
Interpass temperature	375 F max
Postweld heat treatment	1125 to 1150 F/30 h, furnace cool to 600 F
Commercial Weld Wire	
Welding process	submerged arc
Electrode dia	⅛ in.
Flux	type: 1092, size: 20 by 200
Type of power	ac
Current	450 A
Voltage	32 V
Travel	14 in./min
Preheat temperature	250 F
Interpass temperature	350 F max
Postweld heat treatment	1050 to 1075 F/65 h, furnace cool to 600 F; 1075 to 1100 F/15 h, furnace cool to 700 F

Other Mechanical Properties

Tensile Properties

Radiation increases the yield and tensile strength and reduces the uniform strain, the strain hardening exponent, the TS/YS ratio, the elongation (total and uniform), and reduction of area of pressure vessel steels.

Changes in these properties are great for steels irradiated at temperatures below about 450 F as indicated in Figs. 42 and 43 for seven steels (exposed at <250 F) [*33*]. As with notch toughness, however, the response is highly individual. Note the relative increases in yield strength of A302–B versus A350–LF3 (Fig. 43). The very large increases in yield and tensile strength observed with low-temperature high-fluence exposures fortunately are not mirrored by similar large reductions in ductility when the irradiation is conducted at higher temperatures. Temperature effects are indicated by results of testing A302–B steel at various temperatures after irradiation at <250 F. The annealing effects of testing at the higher temperatures are clearly evident in the summary for A302–B in Fig. 44 [*34*]. The effect of temperature upon the critical property, uniform strain, is shown in Table 14 which presents a large volume of data for A302–B steel irradiated at <250 F and at 600 F. The 600 F irradiation affects least the uniform

TABLE 13—*Charpy V-notch ductility characteristics of experimental and commercial weld deposits and of the A543 reference plate before and after irradiation at 550 F to* $\sim 3.5 \times 10^{19}$ *n/cm² > 1 MeV* (*Ref* 30).

Material Identification	Weldment or Plate Thickness, in.	Charpy V-notch 30-ft·lb Transition Temperature, deg F			Approximate Full Shear Energy Absorption Level, ft·lb	
		Unirradiated	Irradiated	ΔT	Unirradiated	Irradiated
Experimental composition 1934	1	−105	−10	95	120	95
Experimental composition 1936	1	−110	−65	45	100	100
Experimental composition 1938	1	− 80	120	200	90	50
Experimental composition 1948	1	− 75	35	110	80	60
Commercial filler (0.24 Cu)	2	−125	290	415	120	57
Commercial filler (0.42 Cu)	2	− 25	505	530	70	34
A543 reference plate	8	−130	−25	105	70	60

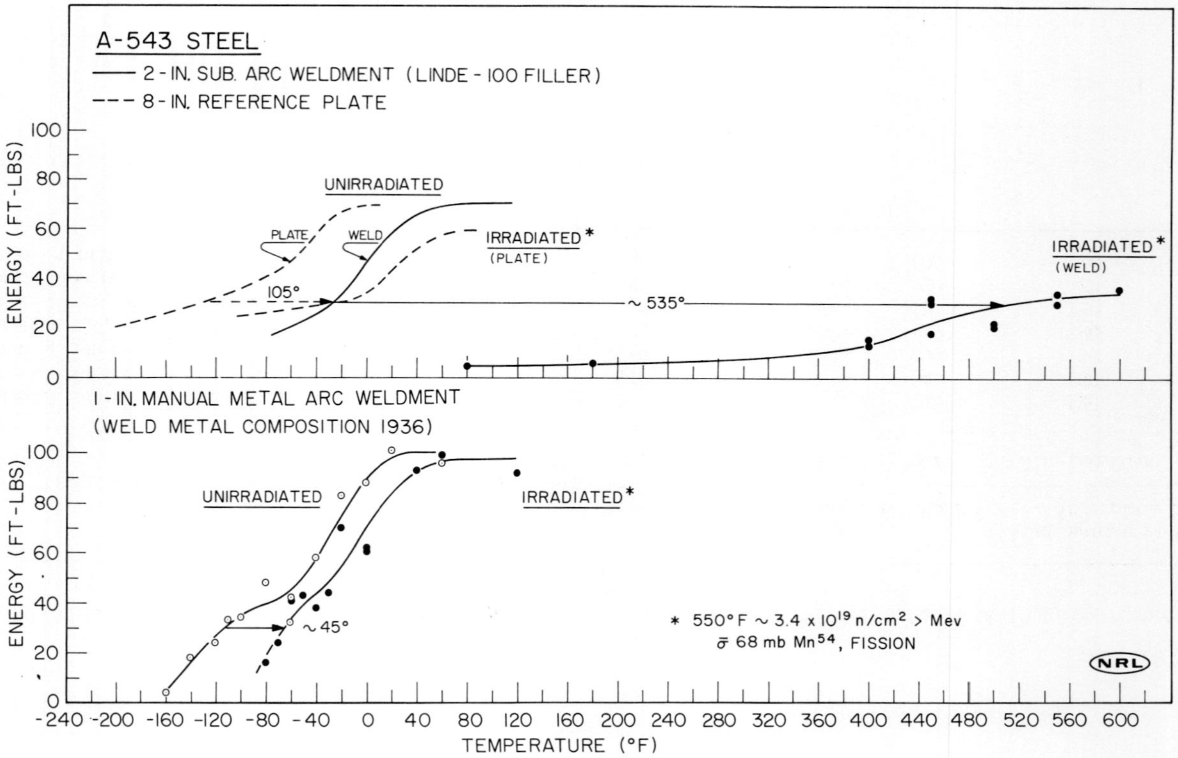

FIG. 41—*Relative resistance of two A543 alternate weld metal compositions to 550 F radiation embrittlement compared to a heavy section A543 reference plate* (*Ref* 31).

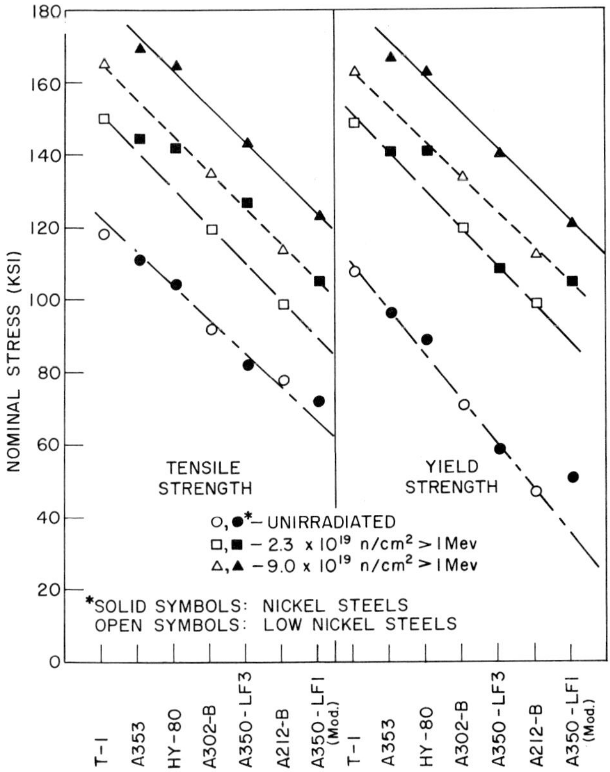

FIG. 42—*The tensile and yield strengths for seven steels after neutron radiation at* < *250 F to the fluence levels indicated* (*Ref* 33).

strain, retaining essentially the same level as the unirradiated steel. These results are quite encouraging since the drastic reduction in uniform strain or elongation is considered most detrimental to the structural integrity of the vessel since it relates the ability (or lack of it) of the vessel to accommodate loads over yield without failure.

The positive aspects of elevated temperature tensile data must be tempered, however, by data of Trozera and Flynn [*35*] in which the uniform strain for tube specimens under equal, positive biaxial stress conditions were greatly reduced from that measured in a solid tensile bar for both irradiated and unirradiated conditions. They state:

> The uniform strain for tubes was typically one fourth that of a solid bar for the same dose (neutron fluence). This implies a large uncertainty in using simple tensile data to predict properties under complicated stress conditions, and a need for more information in this area is suggested.

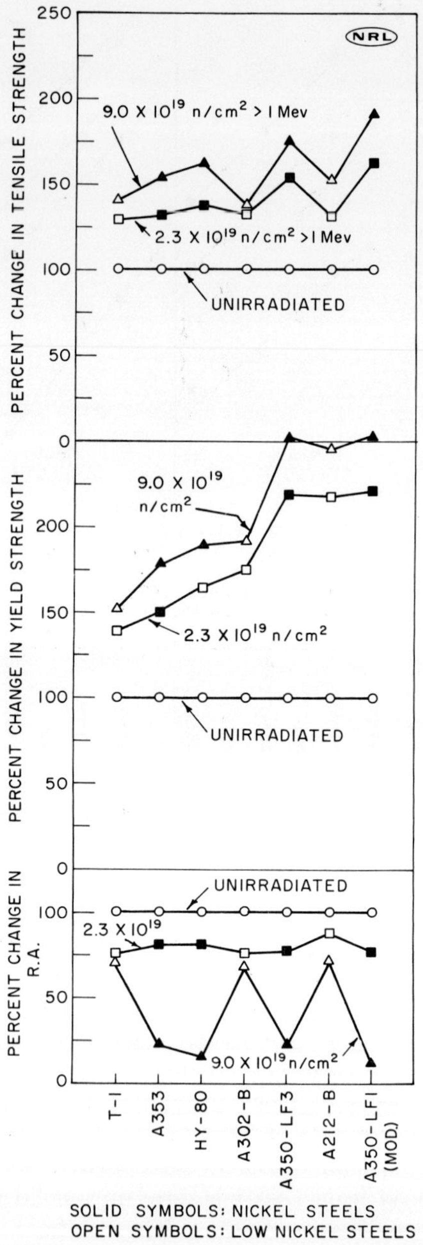

FIG. 43—*The percent change in the tensile and yield strengths and reduction of area with indicated neutron exposures at < 250 F (Ref 33).*

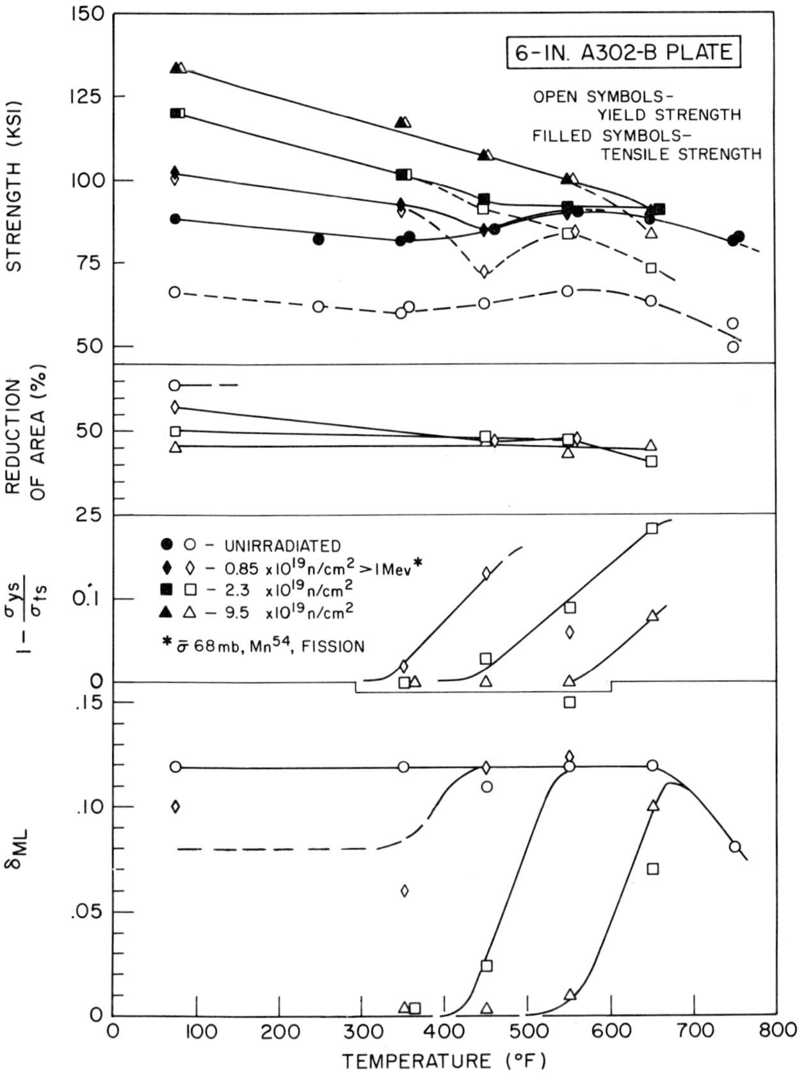

FIG. 44—*The tensile properties of A302-B steel to 750 F for the indicated irradiation conditions* (*Ref* 34).

As fracture mechanics data for irradiated pressure vessel steels (including that for vessel surveillance programs) become more abundant, more and better tensile data also will become available since meaningful fracture mechanics data requires its interpretation relative to the yield strength at the temperature under consideration. Thus, while tensile properties do not appear to be affected seriously by radiation, raising

TABLE 14—*Tensile data for A302–B steel irradiated as indicated* (*Ref* 33).

Specimen Number	Irradiation Temperature, deg F	Neutron Exposure, 10^{19} n/cm² (>1 MeV)	0.2% Yield Strength, ksi	Tensile Strength, ksi	Hardness Rockwell B	Elongation, %	Reduction in Area, %	Natural Strain at Maximum Load
1			68.3	90.6	...	30.5	65.2	0.08
4			68.0	90.8	...	28.2	67.3	0.09
17			68.5	91.7	...	...	63.7	0.11
18			79.0	104.0	...	...	61.7	0.10
20			71.5	94.5	...	24.0	61.2	0.12
23			67.8	90.4	...	27.7	65.6	0.11
24			77.5	1.021	...	27.5	61.2	0.13
25			68.0	91.0	...	30.2	65.5	0.12
27			77.0	100.8	...	28.8	63.1	0.13
28			68.0	90.0	...	26.9	66.0	0.12
Avg			71.4	94.6	...	28.0	64.0	0.12
90	240 (C25–B)	2.0	114.8	116.7	103	11.3	52.6	...
Avg			114.8	116.7	103	11.3	52.6	...
10	240 (C–32)	2.3	122.5	122.6	...	...	50.0	<0.02
22			118.0	118.2	...	...	50.0	<0.02
Avg			120.3	120.4	...	...	50.0	<0.02
11	250 (S–20)	2.2	114.6	116.7	104	...	47.0	<0.03
20			112.4	114.0	102	...	44.0	0.01
Avg			113.5	115.4	103	...	46.0	0.02
60	250 (S–26)	7.0	...	118.5	102	...	...	...
52			118.7	118.7	102	3.0	...	<0.02
Avg			118.7	118.6	102	3.0	...	<0.02
15	200 (S–29)	9.6	133.6	133.8	...	...	44.0	<0.02
12			134.8	135.0	...	...	45.5	<0.02
Avg			134.2	134.4	...	...	45.0	<0.02
53	600 (C–20)	2.3	98.3	112.8	102	...	45.0	0.13
79		5.7	100.6	117.1	103	...	49.0	0.11
Avg			99.5	115.0	103	...	47.0	0.12

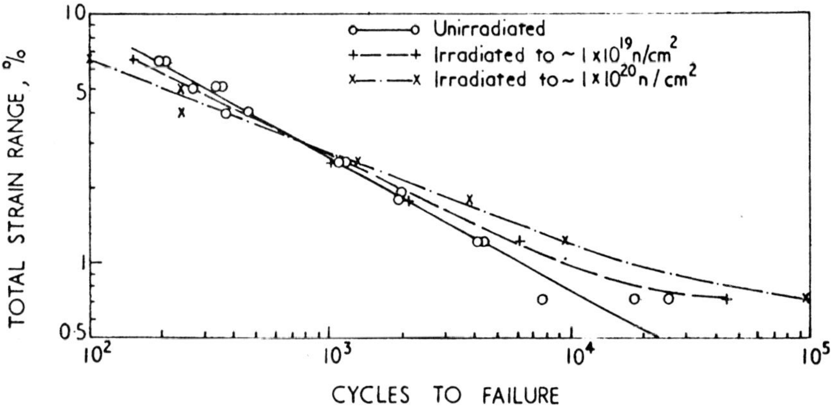

FIG. 45—*Effect of irradiation on the fatigue behavior of a low-copper, high-manganese steel* (*Ref* 37).

questions as to the value of including tension test specimens in vessel surveillance, more sophisticated fracture evaluation techniques will require at least some basic tensile data for irradiated steels.

Fatigue

Vessel surveillance program vessels usually have avoided fatigue specimens (Elk River is a notable exception). Fatigue is of interest to reactor vessel safety primarily because of the possibility of slow crack growth by fatigue to a size critical to brittle fracture. Thus, it is of interest to know how radiation influences both fatigue crack initiation and propagation.

A concise summary of available data on irradiation effects upon the fatigue strength of reactor pressure vessel steels has been provided by Hawthorne and Loss [*36*]. Their summary covers work conducted by six organizations: Bettis Atomic Power Laboratory, Idaho Nuclear Corporation, Reactor Materials Laboratory-Culcheth, England (United Kingdom Atomic Energy Authority), the Southwest Research Institute, the Naval Research Laboratory, and the Centre National d'Recherches Metallurgique, Liege, Belgium.

Though these six laboratories have studied this problem, the experimental effort and the published data are still limited, involving only crack-initiation studies. Nevertheless, the data are quite consistent in suggesting little effect of radiation as indicated by data in Figs. 45 [*37*] and 46 [*36*]. It should be noted, however, that these results do not reflect critical questions of crack-propagation rates in nuclear versus nonnuclear environments. This question is important not only from the point of view of potential failure by fatigue but also for the potential of brittle-fracture crack initiator flaw production. Studies now underway will consider this problem.

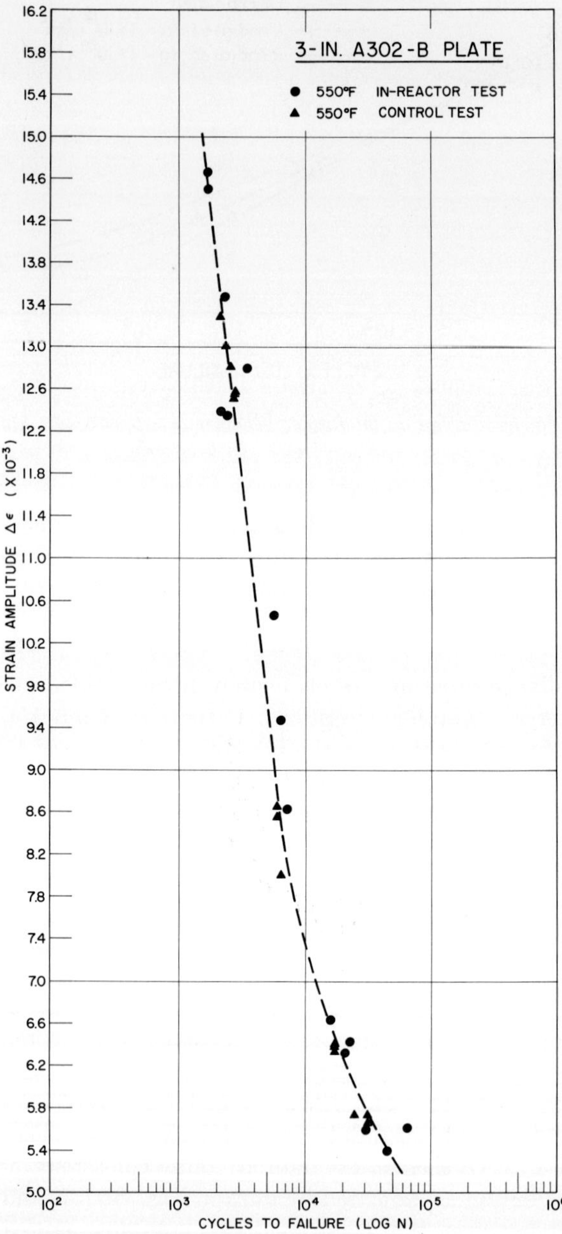

FIG. 46—*Comparison of fatigue data for 3-in. A302-B plate developed by in-reactor tests at 550 F with the results of out-of-reactor control tests. No evidence of an irradiation effect is apparent (Ref 36).*

Summary—(For a more specific summary see covering section, "Synopsis and Highlights")

This review is intended only to provide a basis for comparing neutron embrittlement data from reactor surveillance programs, not a complete analysis of the subject. The plan for this Chapter was to provide a degree of understanding of the factors which affect the level of radiation embrittlement primarily as determined by notch impact data.

The important role of nuclear environment, neutron fluence, flux, and spectrum (to a limited extent) have been reviewed by reference to specific irradiation experiments, usually those performed in test reactors but with comparisons of data from power reactors as well.

Temperature is another important environmental factor which if on the high side works to reduce embrittlement and can be intentionally applied ("annealing") to reduce embrittlement. Other environmental factors, applied stress and hydrogen embrittlement, in conjunction with radiation have been tested and shown not to affect significantly the level of embrittlement. The other important variable in the embrittlement process is the materials effect. This is potentially, if not actually, the most important factor since, though there are very large variations in response steel-to-steel, heat-to-heat, and plate-to-weld-to-HAZ, it has been shown that control of the steel composition and microstructure can produce steels of low or no sensitivity.

Tensile ductility has been shown to be sensitive to the same factors though the degree of change is less severe than changes in fracture toughness. Fatigue strength, on the other hand, is relatively unaffected by irradiation under normal reactor service conditions.

Though relatively abbreviated and admittedly incomplete, it is hoped that a realistic basis for analyzing surveillance data has been provided. The data presented above will be used, at least in general trend fashion, to compare experimental data for radiation embrittlement of steels with surveillance data for radiation embrittlement of specific reactor vessels.

References

[1] Steele, L. E., "Neutron Irradiation Embrittlement of Reactor Pressure Vessel Steels," *Atomic Energy Review*, Vol. 7, No. 2, 1969.

[2] Hawthorne, J. R. and Potapovs, U., "Initial Assessments of Notch Ductility Behavior of A533 Pressure Vessel Steel with Neutron Irradiation," NRL Report 6772, Naval Research Laboratory, 22 Nov. 1968; *Irradiation Effects in Structural Alloys for Thermal and Fast Reactors, ASTM STP 457*, The American Society for Testing and Materials, 1969, pp. 113–134.

[3] Pellini, W. S. and Puzak, P.P., "Practical Considerations in Applying Laboratory Fracture Test Criteria to the Fracture-Safe Design of Pressure Vessels," NRL Report 6030, Naval Research Laboratory, 5 Nov. 1963; *Transactions*, American Society of Mechanial Engineers, Series A, *Journal of Engineering for Power*, Vol. 86, Oct. 1964, pp. 429-443.

[4] Steele, L. E., Hawthorne, J. R., and Gray, R. A. Jr., "Neutron Irradiation Embrittlement of Several Higher Strength Steels," NRL Report 6419, Naval Research Laboratory, 7 Sept. 1966; *The Effects of Radiation on Structural Metals, ASTM STP 426*, American Society for Testing and Materials, 1967 pp. 346–370.

[5] Landerman, E. A., "Yankee Reactor Vessel Surveillance Program Material Property Data," Report PA-PV-177-1, Westinghouse Atomic Power Division, 11 Aug. 1966.

[6] Crider, P. R., Babcock and Wilcox Company, private communication, 25 April 1969.

[7] Chernock, W. P., Combustion Engineering, Inc., private communication, 11 Aug. 1969.

[8] Higgins, J. P., General Electric Atomic Power Equipment Division, private communication, 28 April 1969.

[9] Steele, L. E. and Hawthorne, J. R., "New Information on Neutron Embrittlement and Embrittlement Relief of Reactor Pressure Vessel Steels," NRL Report 6160, Naval Research Laboratory, 6 Oct. 1964; *Flow and Fracture of Metals and Alloys in Nuclear Environments, ASTM STP 380*, American Society for Testing and Materials, 1965, pp. 283–311.

[10] C. Z., Serpan, Jr., and Hawthorne, J. R., "Through-Thickness Notch Ductility and Tension Properties as a Function of Neutron Exposure to a Simulated pressure Vessel Wall of A302-B Steel," NRL Report 6575, Naval Research Laboratory, 30 June 1967; *Nuclear Engineering and Design*, Vol. 6, No. 1 1967, pp. 78–88.

[11] Whitman, G. D., Robinson, G. C. Jr., and Savolainen, A. W., eds., "Technology of Steel Pressure Vessels for Water-Cooled Nuclear Reactors," ORNL-NSIC-21, Oak Ridge National Laboratory, Dec. 1967.

[12] Steele, L. E. and Hawthorne, J. R., "Neutron Embrittlement of Reactor Pressure Vessel Steels," Chapter 16 of *Materials and Fuels for High Temperature Nuclear Energy Applications*, The MIT Press, Cambridge, Mass., 1964, pp. 366–409.

[13] Carpenter, G. F., Knopf, N. R., and Byron, E.S., "Anomalous Embrittling Effects Observed During Irradiation Studies on Pressure Vessel Steels," *Nuclear Science and Engineering*, Vol. 19, 1964, pp. 18–38.

[14] Gray, D. L., "An Effect of Neutron Flux Level Upon Damage Accumulation," USAEC Research and Development Report, HW-61287, Hanford Laboratory, 30 July 1969.

[15] Hinkle, N.E., Ohr, S. M, and Wechsler, M. S., "Dose Rate, Annealing, and Stress Relaxation Studies of Radiation Hardening in Iron," *The Effects of Radiation on Structural Metals, ASTM STP 426*, American Society for Testing and Materials, 1967, pp. 573–592.

[16] Harries, D. R., Barton, P. J., and Wright, S. B., "Effects of Neutron Spectrum and Dose Rate on Radiation Hardening and Embrittlement in Steels," *Radiation Effects on Metals and Neutron Dosimetry, ASTM STP 341*, American Society for Testing and Materials, 1963, pp. 276–293.

[17] Serpan, C. Z ., Jr., and Watson, H. E., "Mechanical Property and Neutron Spectral Analyses of the Big Rock Point Reactor Pressure Vessel," *Nuclear Engineering and Design* (pending publication).

[18] Nichols, R. W. and Harries, D. R., "Brittle Fracture and Irradiation Effects in Ferritic Pressure Vessel Steels," *Radiation Effects on Metals and Neutron Dosimety, ASTM STP 341*, American Society for Testing and Materials, pp. 162–198.

[19] Steele, L. E. et al, "Irradiation Effects on Reactor Structural Materials Quarterly Progress Report, 1 Aug. 31 Oct. 1968," NRL Memoranum Report 1937, Naval Research Laboratory, 15 Nov. 1968.

[20] Serpan, C. Z. Jr., and Hawthorne, J. R., "Yankee Reactor Pressure Vessel Surveillance: Notch Ductility Performance of Vessel Steel and Maximum Service Fluence Determined From Exposure During Cores II, III, and IV," NRL Report 6616, Naval Research Laboratory, 29 Sept. 1967; *Transactions, Journal of Basic Engineering*, American Society of Mechanical Engineers, Vol. 89, No. 4, Series D, Dec. 1967, pp. 877–910.

[21] Reynolds, M. B., "The Effect of Stress on the Radiation Stability of ASTM A302 Grade B Pressure Vessel Steel," *Materials Research & Standards*, Vol. 3, 1963, p. 644.

[22] Hawthorne, J. R. and Loss, F. J., "The Effects of Coupling Nuclear Radiation with Static Service Stresses on Pressure Vessel Material Behavior," *Nuclear Engineering and Design*, Vol. 8, 1968, pp. 108–116.
[23] Harries, D. R. and Broomfield, G. H., "Hydrogen Embrittlement of Steel Pressure Vessel in Pressurized Water Reactor Systems," *Journal of Nuclear Material*, Vol. 9 No. 3, Aug. 1963, pp. 327–338.
[24] Rossin, A. D., Blewitt, T. H., and Troiano, A. R., "Hydrogen Embrittlement in Irradiated Steels," *Nuclear Engineering and Design*, Vol. 4, No. 5, Dec. 1966, pp. 446-458.
[25] Beeston, J. M., Idaho Nuclear Corp. to Steele, L. E., private communication, based on 9th General Meeting of the AEC Irradiation Effects on Reactor Structural Materials Program, Argonne National Laboratory, Ill. Oct. 1968.
[26] Hawthorne, J. R. and Steele, L. E., "Initial Evaluation of Metallurgical Variables as Possible Factors Controlling the Radiation Sensitivity of Structural Steels," NRL Report 6420, Naval Research Laboratory, 29 Sept. 1966; *The Effects of Radiation on Structural Metals, ASTM 426* American Society for Testing and Materials, 1967, pp. 534–578.
[27] Potapovs, U. and Hawthorne, J. R., "Effect of Residual Elements on 550 F Irradiation Response of Selected Pressure Vessel Steels and Weldments," NRL Report 6803, Naval Research Laboratory, 22 Nov. 1968.
[28] Steele, L. E., Knighton, G. W., and Potapovs, U., "Radiation Embrittlement of Pressure Vessels and Procedures for Limiting This Effect in Power Reactors," *Nuclear Applications*, Vol. 4, No. 4, April 1968, pp. 230–244.
[29] Serpan, C. Z., Jr., "Neutron Radiation Embrittlement of LaCrosse Reactor Pressure Vessel Steel and Weldment: Properties and Directionality Consideration," *Nuclear Engineering and Design*, Vol. 8, 1968, pp. 95–107.
[30] Potapovs, U. and Hawthorne, J. R., "The Effect of Residual Elements on 550 F Irradiation Response of Selected Pressure Vessel Steels and Weldments," *Nuclear Applications*, Vol. 5, No. 7, Jan. 1969, pp. 27–46.
[31] Steele, L. E. and Potapovs, U., "Radiation Embrittlement of Reactor Vessel Steels and Suggestions for Its Control," *Nuclear Enginering and Design*, Vol. 8, 1968, pp. 58–70.
[32] Gordon, G. M. and Klepfer, H. H., "Engineering Significance of Ferrite Grain Size on the Radiation Sensitivity of Pressure Vessel Steels," *The Effects of Radiation on Structural Metals, ASTM STP 426*, American Society for Testing and Materials, 1967, pp. 48–66.
[33] Klier E. P., Hawthorne, J. R., and Steele, L. E.,"The Tensile Properties of Selected Steels for Use in Nuclear Reactor Pressure Vessels," NRL Report 6649, Naval Research Laboraty, 26 Dec. 1967.
[34] Hawthorne, J. R. et al, "Irradiation Effects on Reactor Structural Materials, Quarterly Progress Report, 1 May–31 July 1967," NRL Memorandum Report 1808, Naval Research Laboratory, 15, Aug. 1967.
[35] Trozera, T. A. and Flynn, P. W., "Effects of Irradiation On A302-B Steel Subjected to a Multiaxial Stress Distribution," *Flow and Fracture of Metals and Alloys in Nuclear Environments, ASTM STP 380*, American Society for Testing and Materials, 1965, pp. 337–349.
[36] Hawthorne, J. R. and Loss, F. J., "Availability of Data on Irradiated Materials as Related to Design Requirements for Water Cooled Reactor Pressure Vessels," NRL Report 6625, Naval Research Laboratory, 1 Aug. 1967.
[37] Wood, D. S. and Johnson, E. R., "Effect of Irradiation on the High-Strain Fatigue Behavior of a Low-Carbon and a Low-Alloy Steel, "*Journal of the Iron and Steel Institute*, Vol. 205, No. 3, March 1967, p. 305.

CHAPTER 3—ARMY REACTOR VESSEL SURVEILLANCE AND VESSEL EXAMINATION

Introduction	105
SM-1A Reactor	106
Surveillance Provisions	106
SM-1A Vessel Embrittlement Analysis Approach	108
Neutron Flux	109
Embrittlement Trend	112
Annealing Recovery	113
Conclusions on Annealing	119
Discussion	119
Summary	120
SM-1 Reactor	120
Surveillance Provisions and Initial Results	121
Discussion	122
PM-2A Reactor	123
Nondestructive Analysis	124
Prediction of Failure Conditions	124
Pressurization-to-Failure Operations	125
Post-Failure Material Evaluation	127
Notch Ductility and Tensile Properties	127
Fracture Mechanics Properties	131
Summary	134
References	134

CHAPTER 3—Army Reactor Vessel Surveillance and Vessel Examination

Introduction

Among the very earliest power producing reactors were those designed and constructed for the Army. They were built under the Army package power reactor (APPR) concept that dictated small pressure vessels and high flux cores to yield a proportionately high power output for a small physical plant. Because these characteristics result in high fluxes at the pressure vessel, the Army reactors have experienced embrittlement at rates considerably in excess of that exhibited by larger, commercial power reactors. The experience of the Army reactor program can be looked upon as being highly compressed, potential projections of the problems that may be faced by commercial power reactors in the future. The lower operating temperatures and the different, radiation-sensitive steels of the Army reactors accentuate their implications to commercial reactors. For these reasons and the fact that one, the PM-2A, was used as a vessel test experiment for failure analysis, the Army surveillance programs are reviewed here in depth.

Surveillance of Army power reactors was hampered severely by a total lack of specimens exposed at the vessel wall. Nevertheless, the Army effectively worked around this problem by obtaining direct measurement flux values for reactor vessel walls where possible and using them to index embrittlement trends developed from test reactors. As embrittlement became a potentially serious problem, the Army established the precedent of full vessel annealing for notch ductility restoration. The Army reactor characteristics and problems also have prompted development of neutron spectral techniques for correlation of test and power reactor embrittlement data.

A basic premise of the research effort conducted by and for the Army has been to develop or evolve comprehensive trends and criteria that can be transposed from one reactor system to another to meet specific operational problems. Because of this background of experience, as problems arise in commerical power reactors, it may be indeed possible for them to simply obtain minimal data corresponding to a specific problem and use it to enter the analysis scheme available as a result of the Army reactor surveillance programs and associated radiation damage research efforts.

Two important sources of information that have been developed from the Army reactor surveillance programs are the patterns of embrittlement recovery resulting from annealing and the spectral analysis techniques developed to interrelate embrittlement data from diverse reactor systems.

Compounding the problem of high neutron flux in the compact Army reactors are the relatively low operating temperatures which have proven to be conducive to radiation sensitivity. The operating temperatures of the Army reactor vessels have been 510 F or lower (430 F for the SM-1 and SM-1A); these temperatures have the effect of creating radiation damage at a faster rate because the simultaneous correction of atomic displacement damage, common at higher temperatures, is far less efficient in this temperature range. More recent commercial reactors have the advantage of operating above this level.

Another problem of the Army reactors was the sensitivity of the steels to neutron radiation. The SM-1A and PM-2A materials, A350-LF1 (modified) and A350-LF3, showed embrittlement more severe than the general trend of data for a number of steels. Both steels were specified for low initial transition temperature and partially because of the substantial nickel content this objective was met, yielding a 30-ft·lb fix for both forgings at about −80 F. It is only this low initial transition temperature value which permitted large increases for both of these plants. By contrast, the SM-1 reactor vessel of A212-B steel had an initial 30-ft·lb transition temperature of about +50 F; however, the low overall power output of the plant and its low operating history has precluded a large increase in vessel transition temperature.

SM-1A Reactor

The SM-1A reactor is a stationary, medium power (20.2 MWt) pressurized water plant designed as the first model for field emplacement. The plant has a 2 3/8-in.-thick pressure vessel made from A350-LF1 (modified) steel ring forgings. The chemical composition and heat treatment for the vessel steel forgings are given in Table 15. This particular reactor vessel has been studied extensively by the Army Nuclear Power Field Office (NPFO) as well as by NRL which has performed many experimental irradiations on the steel. Accordingly, the response of this steel to neutron radiation is known as well as or better than any other specific reactor vessel steel.

Surveillance Provisions

Because of the compact nature of the SM-1A reactor vessel, it was not possible to position surveillance specimens adjacent to the reactor pressure

TABLE 15—*Chemistry and heat treatment of Army reactor pressure vessel steels.*

Reactor and Type of Steel	Chemical Analysis, weight %										
	C	Mn	Si	P	S	Ni	Cr	Mo	V	Cu	Al
SM–1A, A350–LF1 (modified):											
2 ⅜-in. ring forging	0.13	0.76	0.25	0.023	0.036	1.64	0.16	0.06	0.04	0.08	0.06
3 ⅝-in. test plate	0.15	0.79	0.25	0.027	0.033	1.71	0.05	0.04	0.04	0.088	
SM-1, A212-B:											
4-in. ASTM reference	0.26	0.76	0.24	0.011	0.031	0.22	0.20	0.02	...	...	...
PM-2A, A350-LF3:											
2.4-in. ring forging	0.17	0.53	0.24	0.032	0.03	3.36	...	...	0.035	0.14	...
2.4-in. duplicate forging	0.14	0.52	0.25	0.031	0.032	3.28	0.04	0.05	0.04	...	...

HEAT TREATMENTS

SM-1A—*2 ⅜-in. ring forging and 3 ⅝-in. test plate:* forged in the temperature range of 1700 F min to 2250 F max; normalized at 1600 F; water quenched; tempered at 1250 F; stress relieved at 1150 F.

SM-1—*A212-B, 4-in. ASTM reference steel:* Austenitized at 1650 F for 2 h; water quenched; tempered at 1175 F for 4 h; furnace cooled to below 600 F.

PM-2A—*2.4-in. forging:* forged in the temperature range of 1700 F min to 2250 F max; normalized at 1550 F; water quenched; tempered at 1200 F; stress relieved at 1150 F.

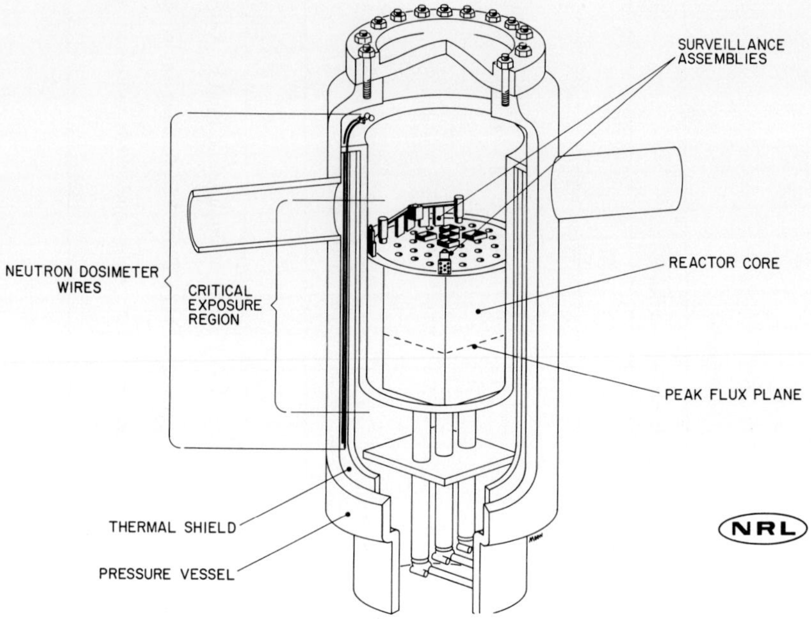

FIG. 47—*Schematic view of the SM-1A rector showing positions of encapsulated surveillance assemblies and neutron dosimeter wires with reference to the core and to the critical pressure vessel regions* (*Ref* 1).

vessel wall. It was possible, however, to place a flux monitor tube alongside the vessel wall as shown in Fig. 47 [*1*].[1] In addition, four capsules containing steel Charpy V-notch specimens representing the SM-1A vessel ring forging and test plate steels were placed at a location above and at the edge of the fuel core. The highest flux intensity >1 MeV at this location was unfortunately only 46 percent of the peak vessel wall flux. Nevertheless, this above-core location has been used by the Army to provide an exposure location for a number of capsules containing SM-1A vessel steel specimens. In fact, certain capsules which were irradiated initially in the low intensity test reactor (LITR) at 430 F were transhipped to the SM-1A reactor where they were installed in the hanger assembly for an additional exposure during the full vessel annealing conducted in the summer of 1967 [*2*].

SM-1A Vessel Embrittlement Analysis Approach

The specific items of data which are of significant value to the analysis of the SM-1A reactor vessel are: (1) the neutron flux measured at the ves-

[1] The italic numbers in brackets refer to the list of references appended to this chapter.

sel wall as modified by a calculated neutron spectrum analysis, (2) the curve relating embrittlement as a function of neutron flux, and (3) the demonstrated recovery of initial mechanical properties after the full vessel annealing.

Neutron Flux

The fast neutron flux measurements made from the exposure of stainless steel tubing placed against the SM-1A vessel wall have been described in detail; the results in terms of an assumed fission spectrum neutron energy distribution are shown schematically with reference to the vessel wall of the SM-1A in Fig. 48 [*3*]. Because the neutron spectrum at the vessel wall is quite different from that of a fission spectrum, as seen in Fig. 49 [*4*], it was necessary to calculate a theoretical vessel spectrum and adjust the fission-spectrum vessel flux to most accurately account for the neutrons actually present. (A detailed discussion of this subject is the topic of Chapter 5 of this report). This adjustment resulted in a full power flux of 7.41×10^{10} n/cm^2/s for neutrons of energies >1 MeV for the pressure vessel wall/stainless steel clad (PVW/SSClad) interface as based upon a calculated neutron spectrum [*3*]. This spectrum adjusted flux measurement for the vessel wall is important since the alternative, a conservatively es-

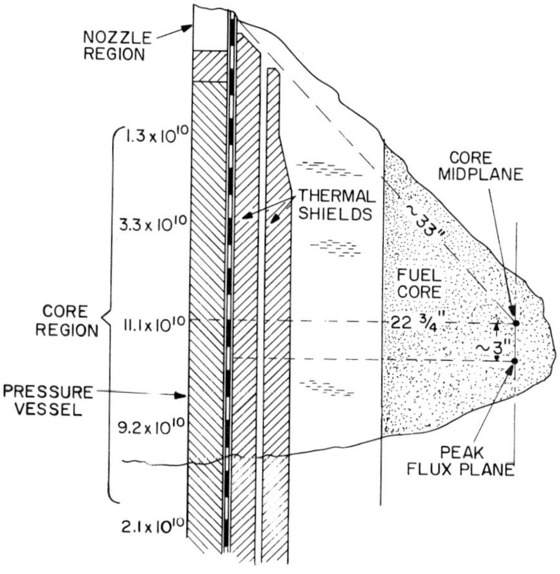

FIG. 48—*Schematic view of the critical pressure vessel region of the SM-1A reactor and the instantaneous neutron flux values measured along the vessel wall. The flux was based upon iron activation in Type 304 stainless steel tubing during reactor operations at an average power level of 8.65 MW* ($\bar{\sigma}$ *68 millibarns (mb), assumed fission spectrum*) (*Ref* 3).

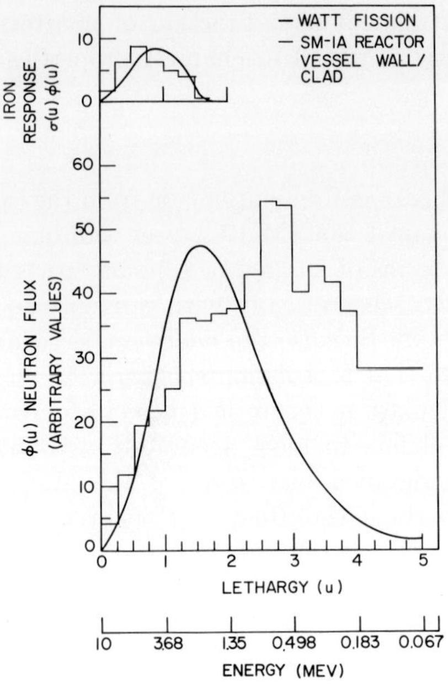

FIG. 49—*Neutron energy spectrum of the SM-1A reactor at the alloy steel/stainless steel clad interface, compared to the Watt fission spectrum (Ref 4).*

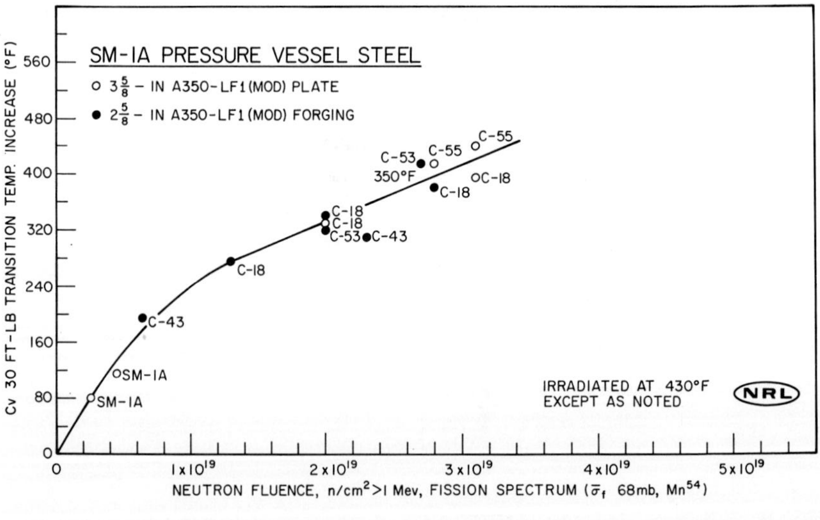

FIG. 50—*Charpy V-notch transition temperature behavior of the SM-1A presure vessel steel irradiated at 430 F. This embrittlement curve will be repeated in subsequent figures (Ref 2).*

TABLE 16—*Charpy V-notch transition-temperature behavior of the SM-1A pressure vessel steel, A350-LF1 (modified), after irradiation at 430 F (Ref 2).*

Material Source	Irradiation Facility	Fluence,[a] n/cm² > 1 MeV	Charpy V-notch 30-ft·lb Transition Temperature, deg F			Full Shear Energy Absorption, ft·lb	
			Initial	Irradiated	ΔT	Initial	Irradiated
SM-1A test plate	SM-1A P.V. (surveillance)	2.6×10^{18}	−45	35	80	...	...
SM-1A test plate	SM-1A P.V. (surveillance)	4.8×10^{18}	−45	70	115	120	115
SM-1A duplicate forging	LITR, C-43	6.4×10^{18}	−80	115	195	130	90
SM-1A duplicate forging	LITR, C-18	1.3×10^{19}	−80	195	275	130	95
SM-1A duplicate forging	LITR, C-53	2.0×10^{19}	−80	240	320	130	90
SM-1A test plate	LITR, C-18	2.0×10^{19}	−45	285	330	120	70
SM-1A duplicate forging	LITR, C-18	2.0×10^{19}	−80	260	340	130	70
SM-1A duplicate forging	LITR, C-53[b]	2.7×10^{19}	−80	335	415	130	75
SM-1A test plate	LITR, C-55	2.8×10^{19}	−45	370	415	120	60
SM-1A duplicate forging	LITR, C-18	2.8×10^{19}	−80	300	380	...	...
SM-1A test plate	LITR, C-18	3.1×10^{19}	−45	350	395	...	...
SM-1A test plate	LITR, C-55	3.1×10^{19}	−45	395	440	120	55
SM-1A duplicate forging	LITR, C-43	2.3×10^{19}	−80	230	310	130	70

[a] $\bar{\sigma}$ 68 millibarns (mb), Mn^{54}, fission.
[b] Irradiated at 350 F.

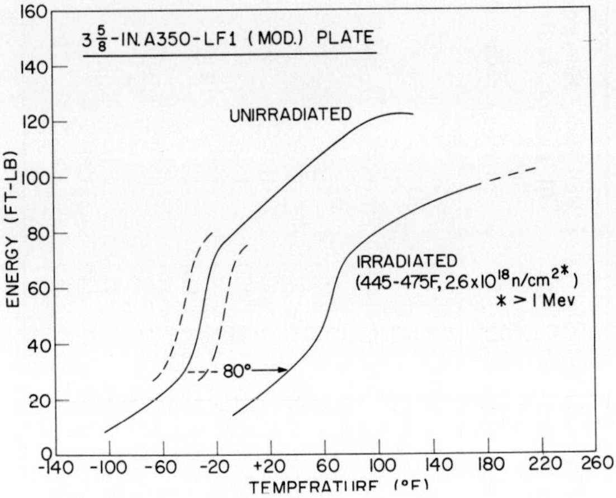

FIG. 51—*Notch ductility characteristics of fabrication test plate (3⅝-in. A350-LFI modified plate) from SM-1A pressure vessel ring forging specimens irradiated in SM-1A above core surveillance position. Dotted lines indicate width of unirradiated data scatterband. Irradiation temperature was determined from low melting alloy monitors (445 F alloy melted, 475 F alloy did not melt)* (*Ref* 5).

timated vessel wall flux value, thus is not required. Further, having measured values permits the confident prediction of meaningful vessel lifetime limits.

Embrittlement Trend

Having a flux value for the vessel wall, the second item for vessel analysis is the behavior trend relating embrittlement and neutron fluence. The basic embrittlement curve for the SM-1A, shown in Fig. 50 [*2*], has been developed with data obtained from the SM-1A surveillance program and with test reactor data as indicated in Table 16 [*2*]. The initial mechanical property surveillance data for the SM-1A is shown in Fig. 51 [*5*]. These data were obtained from specimens in the above-core hanger assembly, the neutron spectrum of which is shown schematically in Fig. 52.

The basic embrittlement curve is repeated in Fig. 53 as a linear projection of the neutron embrittlement that could be expected for the SM-1A vessel wall. Note that there are three temperature scales in the figure. One represents the increase in transition temperature (TT); the other two represent true temperature in degrees Fahrenheit, wherein the transition temperature for plate and forging surveillance steels actually reside as a result of irradiation. These second two scales begin at the initial transition temperature values of −40 and −80 F, respectively, for a 3 5/8-in.-thick fabrication test plate and a 2 5/8-in.-thick duplicate ring forging

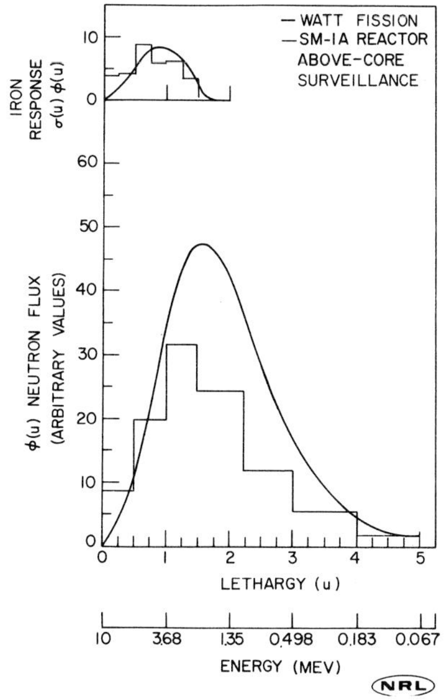

FIG. 52—*Neutron energy spectrum of the SM-1A reactor at the above-core surveillance location, compared to the Watt fission spectrum.*

of A350-LF1 (modified) surveillance steel. We are not concerned at this point with the reannealing behavior only the differences between materials. Note, then, that the duplicate ring forging steel requires a fluence of 2.6×10^{19} n/cm² >1 MeV to reach 300 F, the temperature limit imposed by the Army. It can be seen from Fig. 53, however, that this corresponds to a transition temperature increase of 380 F (because of the initial −80 F transition temperature for the duplicate ring forging). The Army uses the ring forging data for projections of lifetime; this decision is supported by the judgment that the ring forging is more representative of the actual SM-1A vessel material [*2*].

Annealing Recovery

It became clear rather early that embrittlement of the SM-1A vessel would soon be intolerable. The best means for embrittlement relief was shown by a number of studies [*6*] to be postirradiation heat treatment or "annealing."

The conditions established as goals by the Army were one week of annealing at 585 F, as this was the upper limit deemed practical and safe for SM-1A reactor operation using nuclear heat alone. An experimental

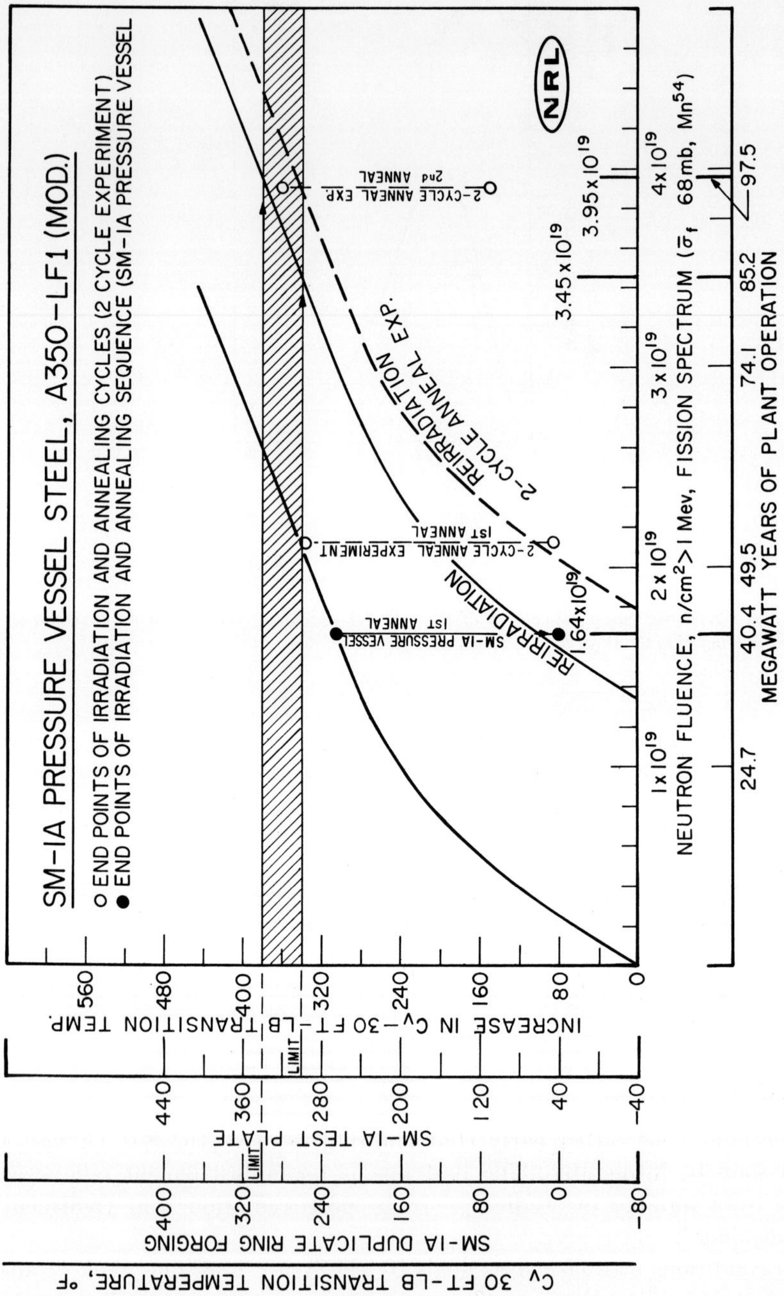

FIG. 53—*Projected reirradiation behavior of the SM-1A pressure vessel following the in-place annealing operation. Fluence values are based on the fission spectrum assumption and a 1-MeV damage threshold (Ref 2).*

investigation was then conducted with these parameters to determine the recovery which could be expected for the SM-1A vessel. The results of two complete cycles of 430 F irradiation/585 F 168-h annealing revealed 70 percent recovery of the initial notch ductility of the ring forging steel [2]. A further study showed that the actual annealing temperature could vary from 560 to 590 F with no significant deviation in the 30-ft·lb value; however, the Charpy upper energy absorption shelf level was restored more readily with the higher annealing temperatures. It is important to recognize here that the 70 percent recovery also means that 30 percent has not been recovered through the annealing treatment. This has been referred to as the residual (~30% of the ΔT). This "residual" embrittlement will occur after each annealing and should be additive. With repeated annealing cycles, such embrittlement may eventually accumulate to a level too high to warrant further annealing. However, such an occurrence would not be expected for the near term (5 to 10 years) operation of the SM-1A.

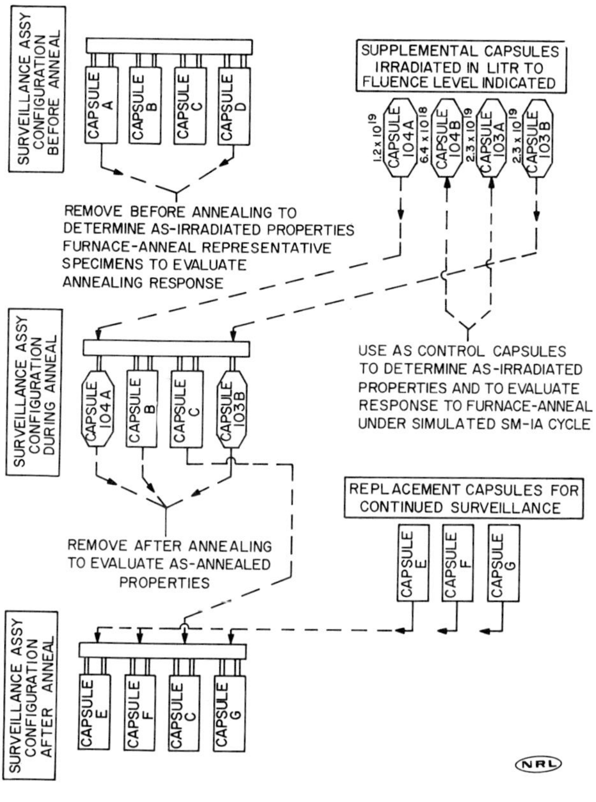

FIG. 54—*The SM-1A surveillance capsule and the supplemental control capsule utilization plan* (*Ref* 2).

With all the conditions for initial embrittlement relief established for the SM-1A reactor vessel, it became apparent that some means were required to prove that meaningful heat treatment of the vessel occurred. This was provided for by utilizing the existing hanger assembly surveillance capsules, along with two new capsules that had been irradiated previously at 430 F in the LITR to match the anticipated SM-1A vessel fluence at the time of annealing. The experimental operations are summarized in Fig. 54.

The actual annealing of the SM-1A reactor vessel was accomplished during the summer of 1967. The conditions achieved were 28 h at 560 F followed by 144 h at 572 F. After the annealing operation the two LITR-irradiated capsules and one of the two remaining SM-1A surveillance capsules were removed from the pressure vessel. The specimens from these three capsules plus the two removed prior to the anneal were used to determine the extent of embrittlement recovery achieved. The results of all five capsules are summarized in Table 17 [*2*] and are shown graphically in Fig. 53.

Figure 55 shows the properties of the SM-1A test plate specimens located in the lower layer of the SM-1A surveillance capsules. The as-irradiated data were taken from the capsule removed prior to the anneal while that for the postannealed condition came from the capsule removed after annealing. A ΔT of 115 F is indicated with a 70 F recovery from the

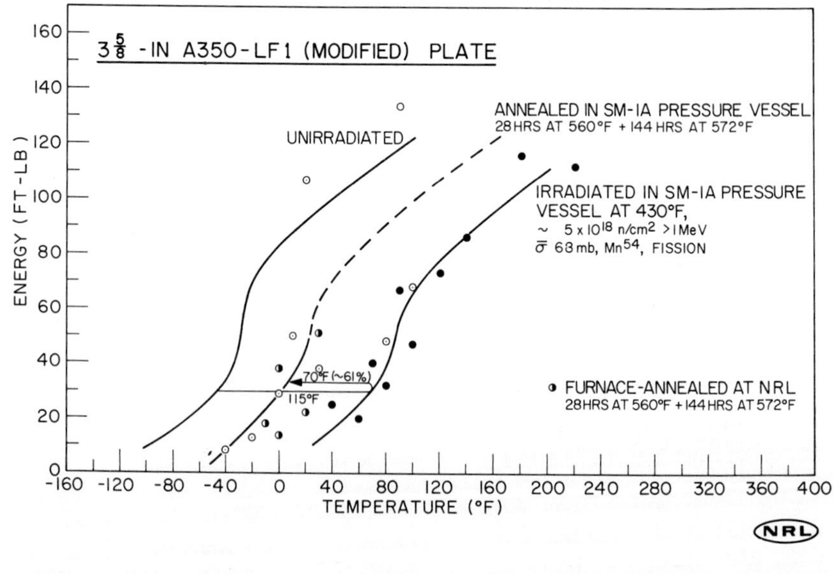

FIG. 55—*Charpy V-notch transition temperature behavior of the SM-1A pressure vessel test plate specimens (lower layer of surveillance capsules) before and after in-reactor annealing* (*Ref* 2).

TABLE 17—*Results from the SM-1A pressure vessel surveillance and the annealing monitor specimen program (Ref 2).*

Specimen Identification	Specimen Material Source	Fluence,[a] n/cm² > 1 MeV	Charpy V-notch 30-ft·lb Transition Temperature, deg F							Recovery, %
			Initial	As-Irradiated	ΔT	Annealed in SM-1A[b]	Furnace Annealed[c]	ΔR	Residual ΔT	
Surveillance capsules A, B, and D (lower specimens)	SM-1A test plate	4.8×10^{18}	−45	70	115	0	...	70	40	61
Surveillance capsules A and D (lower specimens)	SM-1A test plate	4.8×10^{18}	−45	70	115	...	0	70	40	61
Surveillance capsules A, B, and D (center specimens)	4-in. A212-B plate	2.4×10^{18}	10	125	115	105	...	20	95	17
Surveillance capsules A and D (center specimens)	4-in. A212-B plate	2.4×10^{18}	10	125	115	...	60[c]	65	50	57
Surveillance capsule A (upper specimens)	SM-1A test plate	1.2×10^{18}	−45	−20	25	...	...	...	...	...
Surveillance capsules B and D (upper specimens)	SM-1A duplicate forging	1.2×10^{18}	−80	−50	30	−60	...	10	20	33
Supplemental capsule 104A	SM-1A duplicate forging	1.2×10^{19}	−80	180[d]	260[d]	−30	...	210	50	81
Supplemental capsule 104B	SM-1A duplicate forging	6.4×10^{18}	−80	115	195	...	−50	165	30	85
Supplemental capsule 103A	SM-1A duplicate forging	2.3×10^{19}	−80	230	310	20	...	210	100	68
Supplemental capsule 103B	SM-1A duplicate forging	2.3×10^{19}	−80	230	310	...	20	210	100	68

[a] 568 millibarns (mb,) Mn^{54}, fission.
[b] 28 h at 560 F + 144 h at 572 F.
[c] Furnace annealed 48 h at 700 F.
[d] Extrapolated values.

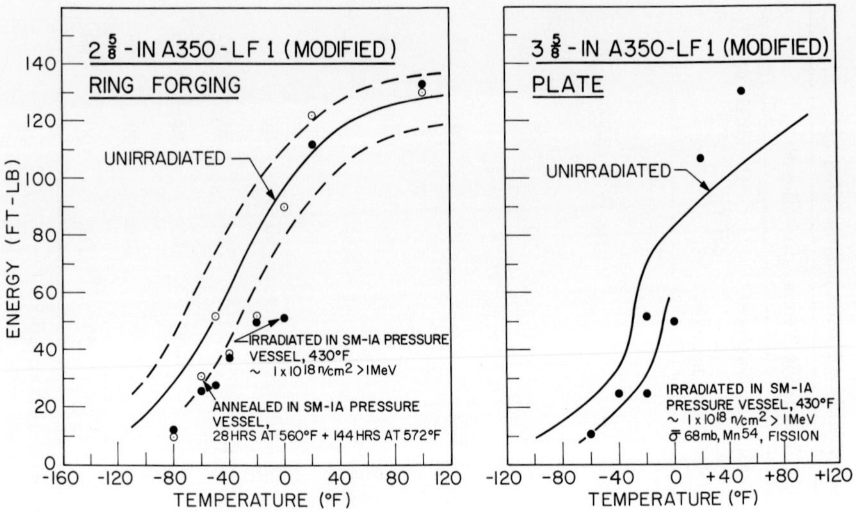

FIG. 56—*Charpy V-notch transition temperature behavior of the SM-1A pressure vessel test plate and the duplicate ring forging specimens (upper layer of surveillance capsules) before and after in-reactor annealing (Ref* 2).

annealing operation. Several of the irradiated specimens were furnace annealed at NRL using a temperature cycle closely approximating that recorded for the actual pressure vessel annealing operation. The annealing response of the furnace-annealed specimens duplicated the recovery behavior of the specimens annealed in the reactor, thus indicating that the recorded SM-1A pressure vessel temperature cycle was realized. The percentage recovery for the SM-1A surveillance-capsule specimens was somewhat lower than that recorded for the LITR irradiated supplemental capsule specimens annealed in the SM-1A (61 percent versus 68 to 81 percent); nevertheless, the residual ΔT was very similar to that recorded for the low fluence LITR supplemental-capsule specimens (Table 16). The as-annealed condition transition temperature for all specimens was only 30 to 50 F higher than the preirradiated-condition transition temperature [*2*].

The results from the SM-1A surveillance assembly specimens of the upper layer in each capsule are shown in Fig. 56. This group contained specimens from both the test plate and duplicate ring forging. Due to the very low fluence at this specimen location, the transition temperature increase was correspondingly small (25 to 30 F); nevertheless, it is significant that the test plate and forging materials responded similarly. Some recovery was indicated due to the pressure vessel annealing operation, but, due to the small initial incremental changes, quantitative estimates are not considered meaningful [*2*].

Conclusions on Annealing

A detailed description and analysis has been published [2]. The main conclusion reached is that induced radiation embrittlement of the surveillance capsule specimens had been reduced between 70 and 80 percent during annealing of the SM-1A reactor vessel. The data also revealed a residual 30-ft·lb Charpy V-notch value of 80 F. The following conclusions, reached after analysis of the annealing surveillance program [2], assumed that the surveillance steel fairly represented the pressure vessel steel.

1. The maximum pressure vessel wall Charpy V-notch 30-ft·lb transition temperature increase before the in-place anneal was 190 F (a 270 F increase over the preservice condition).

2. The maximum pressure vessel wall Charpy V-notch 30-ft·lb transition temperature after the in-place anneal was 0 F (equal to an 80 F residual ΔT value).

3. Based upon Army imposed limits upon vessel NDT and accepting the annealing results, the pressure vessel lifetime following the in-place annealing will be about 5.5×10^{19} n/cm^2 >0.5 MeV (comparable to 124.7-MW years of reactor operation versus a projected 68-MW year lifetime with no annealing).

The most significant conclusion is the projected 124.7-MW year lifetime after anneal versus the 68-MW year lifetime projected without annealing which represents the lifetime increase. If the SM-1A were annealed again, the lifetime would be extended even further.

Discussion

Two important factors require consideration as conditions to the reported annealing and lifetime results for the SM-1A vessel. The first is that a flux value actually has been measured at the vessel wall but that no surveillance specimens were irradiated at the vessel wall in that particular neutron flux. Secondly, the steel used in the surveillance program was not from the vessel forging but rather from a test forging of the same heat as the vessel forgings, formed at the same time using the same practices and heat treatment schedules.

The measured flux value for the vessel wall is quite important because it can eliminate the added margins for conservatism when only calculations are available. The analysis suffers from the lack of embrittlement data on specimens irradiated at the vessel wall. This could be very important since the embrittlement curves drawn for the SM-1A are indeed based primarily upon test reactor data not vessel wall data. It is possible that the vessel wall embrittlement may have been underestimated for the SM-1A reactor. (Possible explanations for this consideration are related to spectrum analysis and are reviewed in a later chapter).

The second critical factor—the vessel steel versus that in the surveillance program—is of concern but for the SM-1A case is not serious based upon present knowledge. The concern here is that the reactor vessel steels have the same sensitivity or response to neutrons as does the surveillance steel. As has been described, there is evidence that microstructure [*7,8*] and composition [*9*] are the primary materials factors responsible for defining the extent of radiation embrittlement. If these are controlled or kept constant, it is reasonable to assume a predictable embrittlement sensitivity within a narrow uncertainty band. In the SM-1A situation, both the vessel and surveillance steel were made from the same heat of steel (eliminating chemistry differences); both were forged and heat treated using the same schedule (hopefully eliminating microstructural differences). Also working toward uniformity between vessel and test steel is the forging operation which usually results in a more uniform microstructure throughout a section than if the material were a rolled plate. It is therefore reasonable to presume that the surveillance steel displays response characteristics equivalent to that of the actual vessel steel.

Summary

The SM-1A reactor surveillance program reveals that much information can be gained from efficient utilization of even limited facilities and materials. Vessel wall dosimetry was used to establish a realistic vessel flux. Test reactor irradiations were used to define embrittlement trends and to establish annealing schedules. Test reactor irradiations were used also to prepare special capsules designed to prove that annealing had taken place within a reactor vessel. Analysis techniques were developed to permit interrelation of test and power reactor data; these also permitted confident projections of the future embrittlement condition of the SM-1A reactor.

SM-1 Reactor

The SM-1 reactor (prototype for the SM-1A) is a stationary, medium power (10 megawatt thermal (MWt) pressurized water plant used by the Army for operating crew training. The 2 5/8-in.-thick vessel is made of A212-B steel roll-bond clad with stainless steel. Chemistry and heat treatment typical for this steel are given in Table 15. The initial drop weight NDT of +40 F was determined by Alco Products, Inc., the reactor builder, on steel trepanned from the vessel [*10*].

Reactor vessel surveillance was truly in its infancy when the SM-1 was in construction, late 1950's, so that radiation damage surveillance was quite limited for this plant. The compact nature of the vessel relative to the core and other internals precluded the inclusion of specimens or flux monitor tubes adjacent to the vessel wall.

TABLE 18—*Increase in brittle-ductile transition temperature for the SM-1 surveillance specimens* (*Ref* 11).

Notch Number[a]	Subsize Izod Transition Temperature Increase, deg F (Charpy V-notch ΔTT values[b] in parentheses)	
	Capsule I 16.4-MW Year	Capsule II 20.8-MW Year
1.............	45 (70)	51 (76)
2.............	58 (83)	67 (92)
3.............	69 (94)	76 (101)
4.............	88 (113)	109 (134)
5.............	120 (145)	141 (166)
6.............	130 (155)	150 (175)

[a] Notches numbered in order of decreasing height (decreasing distance from horizontal core midplane).
[b] Calculated according to ΔTT (Charpy V-notch) = ΔTT (subsize Izod) + 25 F (see Ref *12*).

Surveillance Provisions and Initial Results

Oak Ridge National Laboratory (ORNL) initially placed eight capsules in positions above and to the side of the core. These capsules contained three Izod impact bars each having six notches. Analysis of the early results of these surveillance specimens, presented in Table 18 [*11*], was hampered by a lack of sufficient specimens at a particular location for good curve delineation and by lack of conclusive dosimetry. The subsize Izod values have been adjusted in Table 18 according to a relationship [*12*] which is though to yield approximate Charpy V-notch specimen values. Using more recent neutron flux information [*13*] the Izod impact data were re-evaluated [*14*]. The Izod data points from three samplings were shown to compare favorably [*11, 15*] with Charpy V-notch data for A212-B steel when neutron exposures were calculated from data in Refs *10* and *13*.

After operation of the SM-1 for several years, a location above and to the side of the SM-1 reactor core (similar to that in the SM-1A) was made available by the Army for placement of Charpy V-notch surveillance specimens by NRL. After a 14-month exposure, the two NRL capsules were removed from the above-core location. These capsules contained specimens of ASTM A212-B steel of the reference heat maintained at the Hanford Laboratory. This steel has the same nominal composition as the SM-1 pressure vessel (see Table 15) but has a lower initial 30-ft·lb transition temperature of +5 F. The results of evaluating the upper and lower layers of specimens in the two capsules is shown in Fig. 57 [*5*]. These data

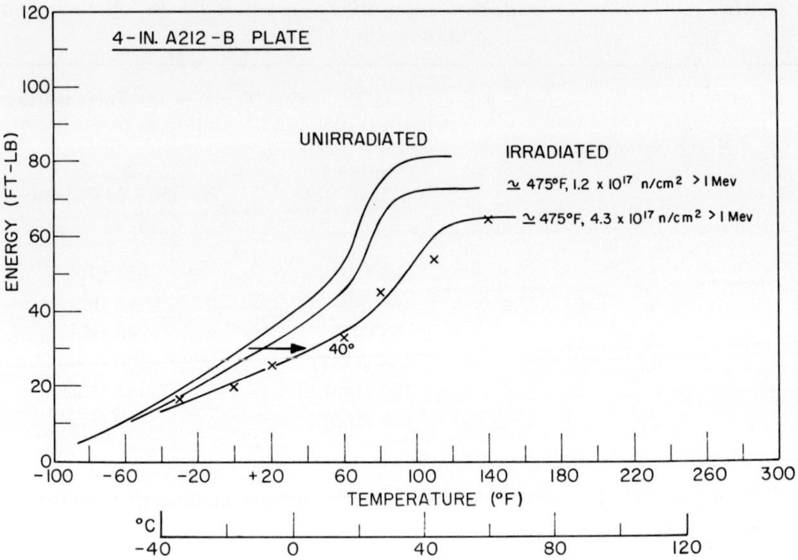

FIG. 57—*Notch ductility characteristics of 4-in. A212-B plate steel representative of the SM-1 pressure vessel. The curve for the 4.3* $\times$ *10^{17} n/cm² exposure was from the bottom SM-1A surveillance capsule section; the lower exposure curve was derived from specimens of the top capsule section* (*Ref* 5).

(plotted in Fig. 58) are found to reinforce trends in transition temperature increase versus neutron exposure suggested by NRL irradiations on the same material but in accelerated test reactor exposures [*16*].

Discussion

In spite of the available comparisons, it is most difficult to draw any substantial conclusions concerning the embrittlement of the SM-1 since for the reasons given above, no measurements of the neutron flux at the vessel wall have been made. The neutron flux is now being carefully determined by NRL in an in-core location, however, and the flux for the inner edge of the vessel wall will be extrapolated therefrom using calculated neutron spectral techniques coupled with spectral refinement using multiple flux monitor foil activation data.

For the present, mock-ups of the SM-1 (and SM-1A) have been made, and the neutron exposures per MW year for the plants have been determined [*10*]. Since the mock-up value for the SM-1A correlates well with the measured value, it is reasonable to assume that the mock-up value for the SM-1 should be a fair approximation as well. The limiting exposure to the SM-1 vessel has been placed by the Army at 9×10^{18} n/cm² >1 MeV to allow for the initial NDT, the "NDT plus 60" safety criterion and

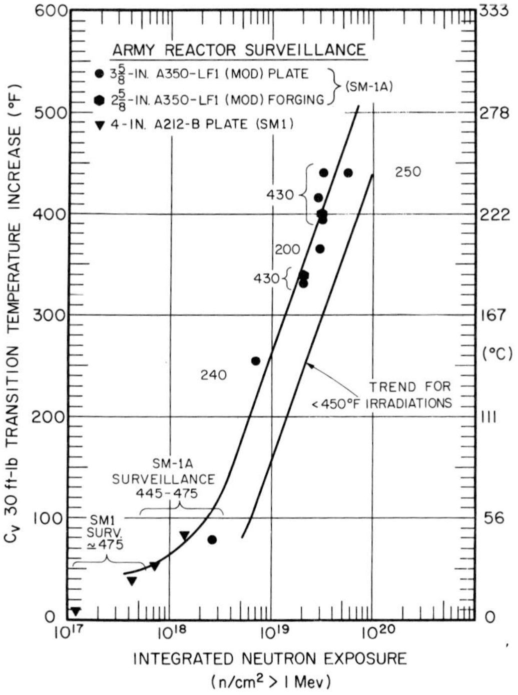

FIG. 58—*Charpy V-notch 30-ft lb transition temperature increase versus integrated neutron exposure for SM-1A and SM-1 pressure vessel steels referenced to <450 F trend band irradiations. Data points near low exposure end of trend band are from respective surveillance programs. Points indicating material behavior at higher exposures are from accelerated irradiations in test reactors. Data for A212-B steel from SM-1A surveillance program are shown for reference* (*Ref* 5).

the radiation-induced transition temperature increase. This mock-up value has indicated to the Army that the vessel life expectancy is about 76-MW years. Since by mid-1969, the SM-1 had reached an accumulated operation equalling 34 MW years (in about 12 calendar years), a continuance of the same type of intermittent, low power, training type operations should permit the reactor to be operated for near its anticipated 20-calendar year lifetime without serious concern for vessel ductility.

PM-2A Reactor

The PM-2A reactor was a portable, medium power (10 MWt) pressurized water plant designated as the second unit for field emplacement and was located within the ice-cap at Camp Century, Greenland. The pressure vessel was fabricated of 2 3/8-in.-thick A350-LF3 ring forgings. The chemistry and heat treatment are given in Table 15. The PM-2A was more compact than either the SM-1 or the SM-1A, and the incident vessel neutron flux at full power was also higher than that for the SM-1A vessel. This higher neutron flux naturally effected neutron embrittlement at a faster rate. The actual vessel flux was not known until removal of the reactor from service so that estimates were necessarily conservative. The embrittlement predicted for the plant, based upon the estimated vessel flux, was great enough to result in restrictive operating conditions on the plant for startup and shutdown.

An Army decision to curtail operations at the Greenland site resulted in the terminal shut down of the PM-2A reactor. This was accomplished on 9 July 1962. The Army and the AEC, recognizing the value of this vessel as a full scale specimen for failure testing, arranged for shipment of the entire reactor vessel assembly to the National Reactor Testing Station in Idaho for these tests.

Nondestructive analysis

The Phase I Analysis portion of the PM-2A reactor test has been reported by Monahan and Halpine [*17*]. In this phase, dimensional, ultrasonic, and liquid penetrant examinations were made to provide data for detailed stress analyses of the vessel. Analytical and nondestructive measurements of the ^{54}Mn residual activity peaks were utilized also to obtain the integrated neutron flux seen by the vessel [*18*]. The primary objective of this phase was to establish existing flaws or out-of-roundness for stress analysis and the level of neutron exposure, in order to project conditions for vessel failure under the controlled conditions of the pressure-temperature test sequence of Phase II. After the nondestructive tests it was not expected that the vessel would fail in the as-irradiated condition, so development of a method for introducing an artificial defect became an important additional portion of the Phase II effort.

Prediction of Failure Conditions

Using available data obtained from the PM-2A vessel shell material, its sister forging, and a model vessel tested to failure, pressures for the full vessel failure were calculated over a range of temperatures for particuar notch geometries. The approach taken by personnel of the Bettis Atomic Power Laboratory (BAPL) who were responsible for the overall PM-2A Phase I and II test direction, was to use fracture mechanics techniques. The complex test sequence involved a series of temperature-pressure levels for a number of flaw conditions. Using the equation for a part-through semielliptical crack and the thin shell formula for hoop stresses, the failure pressure was calculated as a function of temperature, (see Ref *17*). Failure of the PM-2A vessel containing a 7/8-in.-deep, 10-in.-long notch was predicted at a pressure of 4000 $\pm$1000 psig at -20 F [*17*]. Fracture toughness, K_{Ic}, measurements were made by use of wedge

TABLE 19—*Predicted strength and toughness values for two failure temperatures.*

Temperature, deg F	Fracture Toughness, K_{Ic}, ksi$\sqrt{in.}$	Yeild Strength, σ_{ys} ksi
+73	70.0	92.5
−20	57.0	101.0

opening loading (WOL) specimens using the sister forging of PM-2A steel (see Table 15 for chemistry). Of the 15 WOL fracture mechanics specimens cut for use, two orientations with respect to the vessel radius and specimen fracture front were employed. No differences were found, however, due to specimen orientation. Yield strength and fracture toughness values at predicted failure are presented in Ref *17*. These values for the PM-2A sister forging steel are presented in Table 19.

The estimated transition temperature at the vessel shell inside surface was taken as +213 F; the Charpy V-notch energy level "fix" was 35 ft·lb [*17*]. The fluence for the peak exposure location on the vessel inside surface was calculated to be 1.6×10^{19} n/cm^2 >1 MeV from neutron activation measurements of the vessel itself and a P3MG shielding analysis of the neutron spectrum [*18*].

Pressurization-to-Failure Operations

The Phase II portion of the PM-2A vessel test had as its objective the performance of a series of pressurizations at various temperatures and artificial defect sizes covering the region below the reference transition temperature (Charpy energy "fix" transition temperature +60 F), with unstable fracture of the vessel. The complete Phase II testing program has been described in Ref *19*.

Eight full vessel tests were accomplished in the program. They ranged from pressurization up to 5250 psi at −20 F to pressurization to 4000 psi at 272 F for the initial as-irradiated test. Subsequent tests included the introduction of progressively larger artificial defects to the ultimate (test eight) which was 1.75 in. deep and 11 in. long. The defects are indicated in Fig. 59 [*20*]. During Test 6, 250 fatigue cycles were introduced to the vessel by pressurization between 0 and 5000 psig at +72 F. During Test 7, an additional 500 fatigue cycles were accrued under the same temperature and pressure conditions.

Since fatigue cycling had failed to produce fracture of the vessel with the largest defect which could be readily correlated with part-through fracture equations, an alternate crack sharpening procedure was required. Test 8 employed a stress corrosion cracking technique. The procedure consisted of applying a solution of 1:1 hydrochloric acid to the strained

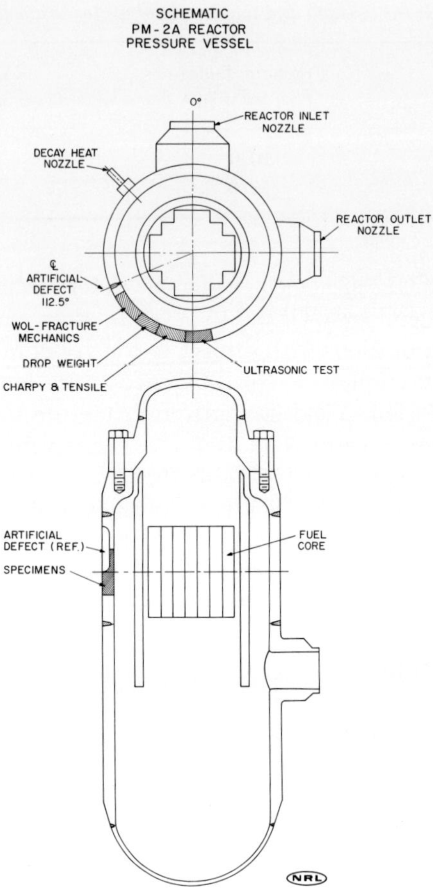

FIG.59—*PM-2A reactor vessel showing the relative locations of the relative locations of the artificial defect, fuel core, and sections of the vessel wall subsequently removed for metallurgical specimen preparation* (*Ref* 20).

vessel to grow a sharp crack about 0.025 in. deep with an end radius less than 0.0001 in. The vessel was treated for 1 h at +72 F under 4000 psig, depressurized, cooled to −20 F, and pressurized to 5250 psig. The vessel was then returned to 70 F, treated with 1:1 hydrochloric acid for 1 h at 5000 psig; it then was depressurized, cooled to −20 F (overnight), and during repressurization it failed in a brittle manner at 4475 psig [*19*].

It is clear that calculation of the PM-2A vessel failure conditions were quite inaccurate in spite of carefully measured flaw, temperature, and stress conditions. This suggests even greater difficulty in computing K_{Ic} (critical crack stress intensity) values for projecting failure or a critical condition in an operating reactor vessel in which all the factors—stress, flaw size, and geometry, temperature, and state of radiation exposure—are less well known than in the PM-2A.

Post-Failure Material Evaluation

Specimen requirements for mechanical properties evaluation were met by the removal of steel from the PM-2A vessel in a horizontal region adjacent to and vertically overlapping the artificial defect (Fig. 59). The fracture mechanics, WOL, specimens were evaluated by Battelle Memorial Institute, Columbus Laboratory; the Charpy V-notch, tension, and drop-weight specimens were evaluated by NRL; and a slab for ultrasonic ("acoustic emission") tests was retained for the Phillips Petroleum Company in Idaho.

Notch Ductility and Tensile Properties

The nil-ductility transition temperature (NDT) of the upper ring forging of the PM-2A vessel was not determined at the time of manufacture. The pre-service Charpy V-notch characteristics taken from the vendor's material certification list (presented schematically in Ref *17*) indicated a 35-ft·lb transition temperature of −58 F. The NDT temperature and 30-ft·lb Charpy V-notch transition temperature were both determined by NRL to be −80 F for the duplicate ring forging. Additional drop-weight specimens machined from the bottom of the lower ring forging of the PM-2A have yielded an NDT of −70 F [*21*]. Since it will be impossible to ever obtain a true "unirradiated" value for the upper ring forging, these two points present strong evidence of an initial −70 F NDT and Charpy

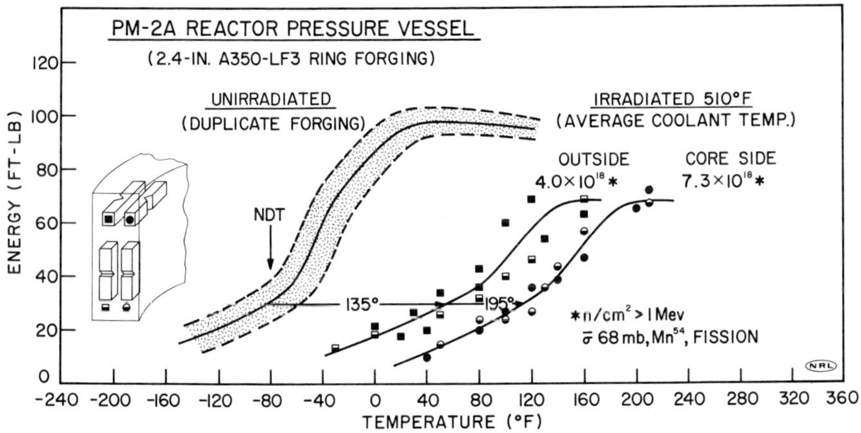

FIG. 60—*Notch ductility characteristics of PM-2A ring-forging steel (A350-LF3). Irradiated specimens were taken from two thickness locations in the ring forging and from both longitudinal and transverse orientations. The unirradiated-condition Charpy V-notch characteristics are those of a sister forging made from the same vessel heat (Ref* 20).

V-notch 30-ft·lb transition temperature for the upper forging of the PM-2A vessel.

The lower ring forging DW specimens used brittle weldbeads for crack starters since the steel was essentially unirradiated. The irradiated upper ring forging DW specimens required the use of pressed-notch crack starters since welding would anneal some of the radiation induced embrittlement thus yielding a tougher condition with lower NDT values. The Charpy V-notch characteristics of the upper ring forging have been developed and are shown with respect to the unirradiated condition of the duplicate forging in Fig. 60 [*20*]. The actual as-irradiated transition temperature for the two vessel thickness locations are shown schematically in Fig. 61 with reference to the corresponding neutron fluences. Note that the transition temperature for the vessel inner edge has been extrapolated from this figure to be +160 F which is considerably less than the initial prediction of 213 F [*17*]. Note also that the fluence extrapolated to the inner edge of the carbon steel vessel wall in 1×10^{19} n/cm^2 >1 MeV. This is confirmed by measurements. This value would be reduced by the fraction 98.26/159,

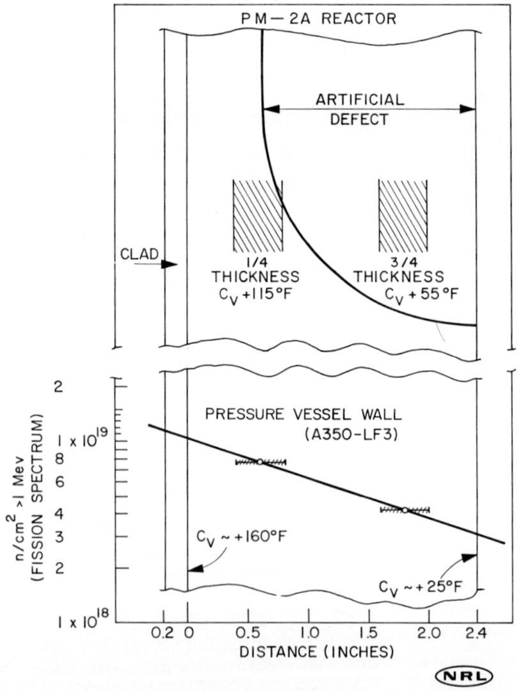

FIG. 61—*Neutron fluence gradient and irradiated-condition Charpy V-notch 30-ft·lb transition temperatures for locations across the thickness of the PM-2A reactor vessel wall. The volume of material occupied by the Charpy V-notch specimens is shown in relation to the artificial defect. The bands on the neutron fluence points indicate the range of material averaged by the measurement* (*Ref* 20).

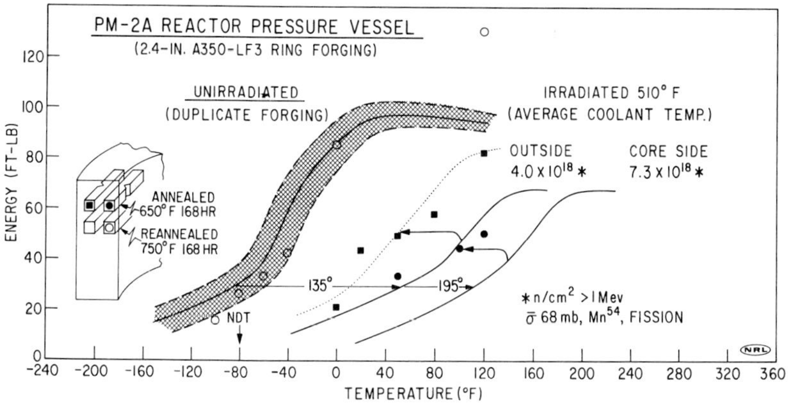

FIG. 62—*Notch ductility characteristics of PM-2A ring-forging steel after postirradiation heat treatment. Closed circles and squares represent specimens heat treated for 168 h at 650 F; recovery of unirradiated-condition properties relative to the sister forging are between 25 to 35 percent. Open circles indicate specimens reheat treated for an additional 168 h at 750 F; recovery of essentially 100 percent of the sister-forging unirradiated-condition properties is indicated* (*Ref* 20).

the ratio of the fission spectrum/calculated spectrum cross sections for n/cm² >1 MeV [*20*]. It is clear that the resulting value is in significant disagreement with the value of 1.59×10^{19} n/cm² >1 MeV initially calculated [*18*]. It must be pointed out, however, that the initial fluence was determined using the P3MG shielding code which is a diffusion-modified code approaching transport theory, while the actually measured value was adjusted on the basis of a pure transport theory code. Perhaps even more significant is the fact that both calculations used different cross-section libraries for all types of scattering, absorption, etc. Accordingly, there can be no direct comparison of the accuracy of one calculation method versus the other except the superficial one relating to the final values determined. This subject is discussed in more detail in Chapter 5.

Several of the irradiated upper ring forging Charpy V-notch specimens were annealed and tested, as shown in Fig. 62. Those annealed for 168 h at 650 F plus 168 h at 750 F show recovery of notch ductility characteristics to a point which is in exceptionally good agreement with the −70 F initial transition temperature deduced above.

Tension specimens of 0.252-in. gage diameter were evaluated at −20 and +72 F. The results of single specimen tests at both temperatures for the two vessel thickness locations are summarized in Table 20 [*20*]. Extrapolation of the yield strengths to the inner edge of the vessel has not been accomplished but would be expected to be in agreement with the yield values or slightly in excess of those predicted in Table 19. The general test behavior for the PM-2A tension specimens consisted of an upper yield

TABLE 20—*Irradiated-condition tensile properties of the pressure vessel wall of the PM–2A reactor (2.4-in. A350–LF3 ring forging) (Ref 20).*

Specimen Location in Vessel	Fluence, n/cm² > 1 MeV	Temperature, deg F	0.2% Yield Point, psi	Tensile Strength, psi	Uniform Elongation, %	Reduction of Area, %	Elongation per 1 in., %	Hardness, R_B
Not irradiated[a]	...	+75	58 600	82 200	...[b]	68.6	33.0	...[b]
¼t	7.3×10^{18}	−20	97 620	111 700	7.0	61.7	...[b]	96
¾t	4.0×10^{18}	−20	90 450	106 100	10.1	67.3	29.8	94.5
¼t	7.3×10^{18}	+72	92 200	105 400	6.2	60.7	24.8	96
¾t	4.0×10^{18}	+72	84 200	99 060	7.8	65.5	28.2	94

[a] Average of six specimens from a sister ring forging.
[b] Not determined.

point rapidly falling to the lower yield point (0.2 percent offset level) which was quite well defined in each case. A comparison of the reduction in area and elongation values of the unirradiated versus the irradiated condition reveals that neither was affected significantly by the irradiation. Both yield and tensile strengths increased markedly, however, thus suggesting that the vessel steel was not damaged significantly with respect to its ability to accept and accommodate the test loading stresses. In fact, with the indicated ductility, the increased yield strength doubtless enhanced its ability to accommodate these stresses. (See Chapter 2 for more general discussion of irradiated steel strength data).

Fracture Mechanics Properties

The results of the PM-2A vessel WOL fracture mechanics tests have been reported and discussed [*22*, *23*]. The results of tests considered valid in terms of proposed ASTM standards are presented in Table 21 [*22*]. Again, note that these values correspond to specimens centered on the 1/4 and 3/4 thickness locations in the vessel and have not been extrapolated to the conditions at the inner edge of the vessel wall. As in the case of the tensile data, greater embrittlement, lower K_{Ic} values, would be projected for this location. However, since the 1/4 thickness of the vessel was located 5/8 in. inside the inner carbon steel edge, a 1-in.-thick WOL specimen would incorporate all but the innermost portion of the vessel steel.

The post-failure fracture mechanics analysis [*23*] has shown that certain

TABLE 21—*Results of fracture toughness studies at −20 F on PM-2A Materials* (*Ref* 22).

Specimen	Crack Orientation	Load at Fracture, psi	Toughness, ksi$\sqrt{\text{in}}$.
	INNER ¼T OF VESSEL, 7.3 × 10[18] N/CM² >1 MEV		
PI-7	Parallel to vessel thickness[a]	5 400	48
PI-9		6 220	67
PI-1T	Parallel to vessel circumference[b]	6 130	65
PI-2T		9 060	91
PI-3T		5 880	48
	OUTER ¾T OF VESSEL, 4.0 × 10[18] N/CM² >1 MEV		
PO-5	Parallel to vessel thickness[a]	9 987	110
PO-6		10 480	112
PO-7		8 100	80

[a] Crack propagation either up or down, vertically, on vessel.
[b] Crack propagation from inner vessel surface through thickness toward outer vessel surface.

meaningful extrapolations of K_{Ic} values can be accomplished for the PM-2A situation. This analysis used the Mode I failure stress equation:

$$\sigma_f = \frac{\Phi K_{Ic}}{\sqrt{1.27\pi a + 0.2017 (K_{Ic}^2/\sigma_y)}} \qquad \text{(1)}$$

where:

σ_f = failure stress in vessel wall remote from the defect (psi),
Φ = elliptical integral of the second kind,
K_{Ic} = Mode I fracture toughness (psi $\sqrt{\text{in.}}$),
a = depth of defect (in.), and
σ_y = yield strength (psi).

The right-hand term in the denominator of the equation above represents a plastic zone correction factor and is discussed in Ref *24*. Utilizing the failure stress equations with various fracture toughness values from Table 21, the actual and computed failure pressures are [*21*]:

	K_{Ic}	Failure Pressure, psi
1. Actual vessel failure pressure	70 500	4475
2. Using K_{Ic} interpolated from "inner" and "outer" specimens	70 500	4200
3. Using K_{Ic} from transverse specimens	74 250	4430
4. Using K_{Ic} from transverse specimens interpolated to the installed defect location	76 500	4650

(Specimens used in 2, 3, and 4 above were fatigue cracked.)

The agreement between the calculated and measured failure pressures shown above is notable. It should be noted, however, that initial projections of the vessel failure stress for the full vessel were significantly in error: 5800 psi projected versus 4475 actual. The agreement for Lines 4 and 1 is valid because the crack tips for both the transverse specimens and the vessel steel defect were nearly the same depth into the inner edge of the vessel wall. Hence, crack-tip embrittlement in defect and specimen should have been very similar.

Although crack initiation is of high importance to fracture mechanics, so also is the toughness of the material behind the crack. For the vessel, the steel was embrittled more as the crack ran in toward the reactor core; while for the specimen, the crack ran away from the reactor core into less highly embrittled steel. The implications of this difference to the PM-2A analysis are not clear.

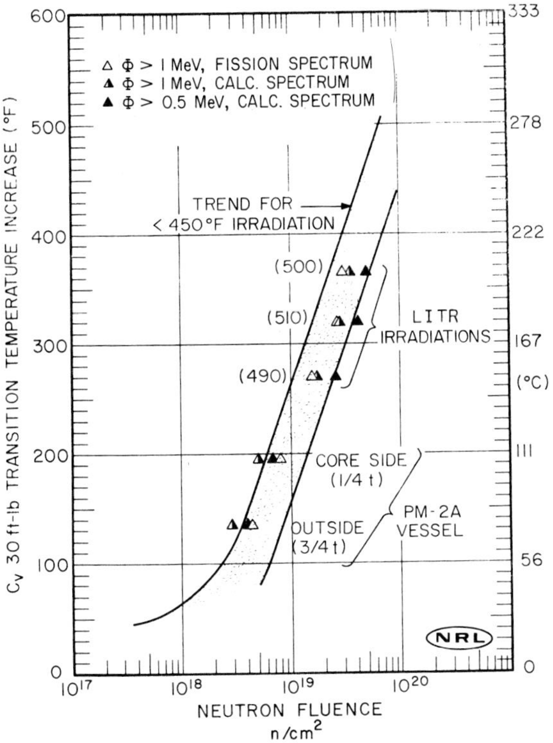

FIG. 63—*Transition-temperature increase (indexed to Charpy V-notch 30-ft·lb level for a sister ring forging) of the PM-2A reactor pressure vessel A350-LF3 forging steel versus neutron fluence determined by assuming a fission spectrum (open triangles) and calculated spectra (partially and completely closed triangles). The LITR data were from irradiations at the indicated temperatures using the sister forging as specimen material* (*Ref* 20).

One of the principal goals of the PM-2A study was to compare the notch ductility data of the irradiated vessel against damage trends developed in laboratory tests and accelerated irradiation experiments. Such a comparison has been made [*20*] and is summarized below. Data for several LITR irradiations of PM-2A duplicate forging steel conducted at 490 to 510 F are shown with the PM-2A vessel data in Fig. 63 [*20*]. This figure reveals that data plotted on the basis of a fission spectrum correlate well. However, when the data are plotted on the basis of transport theory computed neutron spectra for all irradiation locations concerned, the solid triangles are seen to form two different damage trends that describe the test and the power reactors. There is no explanation for this apparent anomaly at this time, although an intensive study is being carried on to uncover the causes.

Summary

The PM-2A experience has aided fracture mechanics analysis techniques because it demonstrated validity of certain equations and concepts when the fracture parameters are well known. However, it must be noted that both the pre-failure K_{Ic} prediction and post-failure fracture toughness estimates were, respectively, too low and too high by comparison with the values measured from vessel steel specimens.[2]

The steel in the pressure vessel wall was significantly embrittled although tensile ductility was not greatly affected. The increases in ductile-brittle transition temperature for steel in the pressure vessel compare well with trends developed in test reactors only on the basis of an assumed fission spectrum neutron distribution.

Neutron dosimetry data from locations across the thickness of the vessel wall present a linear relationship when plotted on a semilog scale. Therefore, it should be possible to accurately estimate through-thickness embrittlement and neutron dosimetry in pressure vessels of reactors provided a valid dosimetry value can be obtained from the vessel itself or from surveillance data.

Greater stresses were required to fracture the vessel than would be encountered during normal service. Even with this higher stress, a very large flaw also was required to effect the PM-2A fracture. Accordingly, the fracture experience with the PM-2A reactor suggests increased confidence in the overall safety and reliability of reactor vessels.

It should be noted in caution, however, that the PM-2A vessel was made up of forged rings with no welds in the high flux region of the vessel, and thus the favorable conclusions stated above cannot be translated directly to welded vessels where materials discontinuities may affect both irradiation insensitivity and crack initiation.

References

[*1*] Potapovs, U., Knighton, G. W., and Denton, A. S., "Critique of In-Place Annealing of SM-1A Nuclear Reactor Vessel," *Nuclear Engineering and Design*, Vol. 8, 1968, pp. 39–57.

[*2*] Potapovs, U., Hawthorne, J. R., and Serpan, C. Z., Jr., "Notch Ductility Properties of SM-1A Reactor Pressure Vessel Following the In-Place Annealing Operation," NRL Report 6721, Naval Research Laboratory, 21 May 1968.

[2] For surveillance considerations, the use of fracture mechanics specimens may add better knowledge of K_{Ic} for the specific condition at the surveillance position, but this can only provide some general guidance such as flaw size limits to be imposed in the vessel. A further disadvantage of fracture mechanics specimens (of sizes suitable for surveillance) is the very low temperatures required for producing valid K_{Ic} values even after relatively high neutron fluences. Thus, it is expected that much experience in the use of fracture mechanics techniques for the tough pressure vessel steels will be required before routine and meaningful application in reactor vessel surveillance programs is possible.

[3] Serpan, C. Z., Jr., and Steele, L. E., "Neutron Spectral Considerations Affecting Projected Estimates of Radiation Embrittlement of the Army SM-1A Reactor Pressure Vessel," NRL Report 6474, Naval Research Laboratory, 30 Sept. 1966.

[4] Potapovs, U., Hawthorne, J. R., and Serpan, C. Z., Jr., "Notch Ductility Properties of SM-1A Reactor Pressure Vessel Following the In-Place Annealing Operation," *Nuclear Application*, Vol. 5, No. 6, Dec. 1968, pp. 389–409.

[5] Serpan, C. Z., Jr., Steele, L. E., and Hawthorne, J. R., "Radiation Damage Surveillance of Power Reactor Pressure Vessels," NRL Report 6349, Naval Research Laboratory, 31 Jan. 1966; *Transactions*, American Society of Mechanical Engineers, Vol. 89, Series D, No. 1, March 1967, pp. 221–232.

[6] Steele, L. E., Hawthorne, J. R., and Gray, R. A., Jr., "Neutron Irradiation Embrittlement of Several Higher Strength Steels," NRL Report 6419, Naval Research Laboratory, 7 Sept. 1966; *The Effects of Radiation on Structural Metals, ASTM STP 426*, American Society for Testing and Materials, 1967, pp. 346–370.

[7] Carpenter, G. F., Knopf, N. R., and Byron, E. S., "Anomalous Embrittling Effects Observed During Irradiation Studies on Pressure Vessel Steels," *Nuclear Science and Engineering*, Vol. 19, 1964, pp. 18–38.

[8] Hawthorne, J. R. and Loss, F. J., "The Effects of Coupling Nuclear Radiation with Static and Cyclic Service Stresses and of Periodic Proof Testing on Pressure Vessel Materials," NRL Report 6620, Naval Research Laboratory, 1 Aug. 1967.

[9] Potapovs, U. and Hawthorne, J. R., "The Effect of Residual Elements on 550 F Irradiation Response of Selected Pressure Vessel Steels and Weldments," NRL Report 6803, Naval Research Laboratory, 22 Nov. 1968; *Nuclear Application*, Vol. 5, No. 7, Jan. 1969 pp. 27–46.

[10] McLaughlin, D. W. et al, "Effect of Radiation Damage on SM-1, SM-1A, and PM-2A Reactor Vessels," APAE No. 107, Alco Products, Inc., 14 Oct. 1961.

[11] Berggren, R. G. et al, "Impact Tests on Irradiated Pressure-Vessel Steels," ORNL-3364, Oak Ridge National Laboratory, Solid State Division, Annual Progress Report, 31 Aug. 1964, pp. 147–149.

[12] Berggren, R. G., "Radiation Effects in Ferritic Steels," TID-7588, *USAEC Conference on the Status of Radiation Effects Research on Structural Materials and the Implications to Reactor Design*, 15–16 Oct. 1969, pp. 91.127.

[13] Moote, G. F., Walthousen, L. D., and Clancy, E. F., "Fast Neutron Flux in a Mockup of SM-1A Core and Vessel," ACNP-63586, Allis-Chalmers Mfg Co., Aug. 1963.

[14] Berggren, R. G. and Schaffer, L. D., "Army Reactors Program Annual Progress Report," 31 Oct. 1963.

[15] Berggren, R. G. et al, "Gas Cooled Reactor Program Semi-Annual Progress Report for Period Ending March 31, 1965," ORNL-3807, Oak Ridge National Laboratory, pp. 264–266.

[16] Steele, L. E. et al, "Irradiation Effects on Reactor Structural Materials, Quarterly Progress Report, 1 May–31 July 1965," NRL Memorandum Report 1638, Naval Research Laboratory, 15 Aug. 1965.

[17] Monahan, J. M. and Halpine, P. A. "PM-2A Reactor Vessel Test—Phase I," WAPD-TM-604, Westinghouse Atomic Power Laboratory, July 1966.

[18] Shure, K. and Oberg, C. T., "Neutron Exposure of the PM-2A Reactor Vessel," *Nuclear Science and Engineering*, Vol. 27, 1967, pp. 348–359.

[19] Monahan, J. M. and Walker, T. J., "PM-2A Reactor Vessel Test–Description of Testing and Failure Conditions," WAPD-TM-640, Westinghouse Atomic Power Laboratory, Jan. 1967.

[20] Serpan, C. Z., Jr., "Notch Ductility and Tensile Property Evaluation of the PM-2A Reactor Pressure Vessel," NRL Report 6739, Naval Research Labortory, 19 June 1968.

[21] Serpan, C. Z., Jr., "Test of PM-2 A Reactor Vessel, Project 10205," *Nuclear Safety*, Vol. 10, No. 1, Jan.–Feb. 1969.

[22] Claudson, T. T., Williams, J. A., and DeMastry, J. A., "Fracture Toughness of Irradiated and Unirradiated Heavy Section Pressure Vessel Material," ASME Paper 68-WA-ER-2, Winter Annual Meeting, 1–5 Dec. 1968.

[23] Walker, T. J., "PM-2A Reactor Vessel Test–Analysis of Failure Conditions," WAPD-T-198, Westinghouse Atomic Power Laboratory, March 1968.

[24] *Fracture Toughness Testing and Its Applications, ASTM STP 381*, American Society for Testing and Materials, 1965.

CHAPTER 4—COMMERCIAL REACTOR VESSEL SURVEILLANCE

Introduction	139
Big Rock Point Reactor	145
Surveillance Program	145
Notch Ductility Properties	145
Base Metal	146
Heat Affected Zone	147
Weld Metal	149
Vessel Lifetime Projections	153
Neutron Dosimetry and Spectrum Considerations	153
Projections	154
Summary	156
Yankee-Rowe Reactor	158
Surveillance Program	158
Notch Ductility Properties	158
Accelerated Capsule Results	158
Vessel Wall Capsule Results	162
Discussion	163
Vessel Lifetime Projections	168
Embrittlement Relief	171
Through Thickness Embrittlement Evaluation	171
Summary	173
Dresden Nuclear Power Station	175
Surveillance Program	175
Notch Ductility Properties	179
Tensile Properties	183
Vessel Lifetime Projections	183
Neutron Dosimetry and Spectrum Considerations	183
Projections	187
Summary	188

Elk River Reactor 189
Surveillance Program 189
Materials 189
Surveillance Capsules and Locations 191
Notch Ductility Properties 193
Embrittlement Projection 194
Tensile Properties 197
Fatigue Properties 198
Summary 200
Humboldt Bay Reactor 200
Surveillance Program 200
Notch Ductility Properties 202
Tensile Properties 203
Summary 204
References 205

CHAPTER 4—Commercial Reactor Vessel Surveillance

Introduction

In spite of the large number of nuclear plants in the general category of operating, under construction, or planned, few had produced surveillance data by the end of 1968. Many programs are planned, however, and for perspective it may be of value to consider the reactors for which data on anticipated increase in NDT and planned surveillance are available. The small number of operating plants for which surveillance programs are now available limit severely the degree of specificity in tabulations of descriptive surveillance data.

The first step toward planning a surveillance program is usually the analysis for anticipated vessel exposure and the projection of NDT increases which the vessel exposure can be expected to produce. In some cases, of course, uncertainty as to exposure or past experience with surveillance programs may determine the nature of the surveillance program. Data on both anticipated NDT increase and actual or planned surveillance programs have been compiled [*1*][1] and form the basis for the Tables 22 and 23 which seek to summarize data on both anticipated exposure with NDT increase and surveillance program plans. (It is difficult to summarize in a table plans for a number of individual plants, but, for comparisons sake, it is worthwhile even if the result is an incomplete presentation).

Table 22 outlines projections for those reactors reporting in detail. Table 23 is supplementary, providing available data on those plants for which detailed results are not available. It must be remembered in comparing projected fluences and NDT values that much of the data are calculations or best projections and operational results may vary considerably. Even so, the fluence-NDT data suggest for some reactors relatively serious levels. In terms of planned programs, it should be noted that for the early reactors such as Dresden and Yankee-Rowe, the surveillance programs were developmental leading to later more standarized programs. Also,

[1] The italic numbers in brackets refer to the list of references appended to this chapter.

TABLE 22—*Estimation of radiation damage and summary of pressure vessel materials surveillance programs in U.S. water reactors (Ref 1).*

Reactor	Pressure Vessel Steel	Reactor Life, years	Assumed Thermal Power, MW	Assumed Plant Factor	Maximum[a] Lifetime Fluence, n/cm² > 1.0 MeV	Projected Maximum NDT		Surveillance Material[b]	Specimens at Vessel Wall		Other Specimens	
						deg F	deg C		Tensile	Charpy	Type	No.
Dresden I.......	A302–B	40	626	0.8	2.5×10^{19}	220	104	V, W, HAZ and GE Ref[c]	42	234	...[d]	
Yankee–Rowe...	A302–B	30	600	0.85	2.5×10^{19}	315	157	V (1 heat) and Ref	4	36 and 8[e]	...[d]	
Saxton.........	A212–B[f]	5	23	0.60	2.0×10^{19}	250	121	V(1 heat) and Ref	40	192[g]		
Big Rock Point..	A302B	40	157	1.0	3.63×10^{18}	60	16	V, W, HAZ (3 plates) and GE Ref	30 and 4	135 and 21[e]	...[d]	
Elk River.......	A302–B	20	51	0.80	1.9×10^{19}	340	171	[h] and Ref			fatigue	156[d]
Humboldt Bay..	A302–B	20	163	1.0	3.4×10^{18}	70	21	V, W, HAZ (3 plates) and Ref	30 and 4	135 and 21[e]	...[d]	
BONUS........	A212–B	20	50	1.0	1.65×10^{18}	270	132	V, W, HAZ		60	...[d]	
Pathfinder......	A212–B	20	200		1.14×10^{18}	90	32	V and Ref	14	40		
LaCrosse.......	A302–B	20		0.80	9.5×10^{18}			V and Ref			...[d]	
San Onofre......	A302–B	30	1347	0.9	6.1×10^{19}	330[i]	166	V (3 plates), W, HAZ (1 plate) and Ref	40	192	WOL	48

Connecticut Yankee.......	A302–B	30	1473	0.9	2.5×10^{19}	255[i]	124	V (3 plates), W, HAZ (1 plate) and Ref	40	192	WOL	48
Brookwood......	A302–B	40	1300	0.8	2.2×10^{19}	333	167	V (2 forgings), W, HAZ (1 forging) and Ref	54	288	WOL	54
Indian Point 2..	A533–B	40	2758	0.8	0.9×10^{19}	178[i]	81	V (3 plates), W, HAZ (1 plate) and Ref	27	256	WOL	42
Palisades I......	A302–B, mod,[j]	40	2450	0.80	9.4×10^{18}	215	102	V, W, HAZ	...[k]	...[k]		
Turkey Point 3..	A508II[l]	40	2300	0.8	2.6×10^{19}	260[i]	127	V (2 forgings), W, HAZ (1 forging) and Ref	27	256	WOL	42

[a] Computed or estimated values.

[b] V = pressure vessel vase plate; W = weld pressure vessel plate; HAZ = heat affected zone of weld; Ref = ASTM reference A302–B plate; GE Ref = GE reference steel from Humboldt Bay vessel stock.

[c] Weld and HAZ specimens were added in late 1962.

[d] Also included a number of C_v and tension specimens at accelerated irradiation positions.

[e] Number of reference steel specimens.

[f] Multilayer pressure vessel.

[g] Subsize specimens from multilayer vessel material.

[h] Steel of composition similar to that of pressure vessel.

[i] Initial NDT not given, assumed initial NDT = 10 F; 100 F for Brookwood.

[j] Nickel content of 0.4 to 0.7 percent (same as A533–B).

[k] Program not firm.

[l] Forgings.

TABLE 23—*Estimation of radiation damage of vessels in various U. S. water reactors* (*Ref* 1).

Reactor	Pressure Vessel Steel	Reactor Life, years	Assumed Thermal Power, MW	Assumed Plant Factor	Estimated Maximum Lifetime Fluence, $n/cm^2 > 1.0$ MeV	Projected Maximum NDT		Surveillance Information[a]
						deg F	deg C	
Shippingport	A302–B	20	225	not given	6.0×10^{19}	438	226	unavailable
Indian Point I	A212–B	40	615	0.8	7.5×10^{18}	290	143	no program
Oyster Creek I	A302–B, Mod[b]	40	1860	1.0	9.2×10^{17}	80	27	standard program[c]
Nine-Mile Point	A302–B, Mod[b]		1846	1.0	7.26×10^{17}	70	21	standard program[c]
Dresden 2	A302–B, Mod[b]	40	2467	1.0	2.4×10^{17}	60	16	expanded program[d]
Millstone Point	A302–B, Mod[b]		2011	1.0	4.9×10^{17}	80	27	standard program[c]
Brown's Ferry I and II	A302–B, Mod[b]	40	3293	1.0	3.5×10^{17}	80	27	special program[e]
Dresden 3	A302–B, Mod[b]	40	2467	1.0	$<1.0 \times 10^{19}$	60	16	expanded program[d]
Quad Cities I and II	A302–B, Mod[b]	40	2527	1.0	2.4×10^{17}	60	16	expanded program[d]
Monticello NGP	A302–B, Mod[b]	40	1674	1.0	5.4×10^{17}	90	32	standard program[c]
Vermont Yankee	A302–B, Mod[b]	40	1593	1.0	1.0×10^{17}	40	4	standard program[c]

[a] Among the other announced reactors, Boston Edison, Hartsville, Oconee 1, 2, and 3, Point Beach and Turkey Point 4 have planned surveillance programs, but no detailed information is available.

[b] Nickel content of 0.4 to 0.7 percent (same as A533–B).

[c] See text for general description of GE and Westinghouse surveillance programs.

[d] GE expanded program includes standard program plus accelerated program including electroslag and submerged arc welds plus related HAZ.

[e] Limited to electroslag.

surveillance capsule and program design was in its infancy; hence, early programs (which will be described in succeeding pages) are not to be considered representative.

In the critique of surveillance programs, Chapter 6, comparisons of projections and measured results will be provided. It is clear from the comparisons of Tables 22 and 23 that there is a vast divergence of projected fluences and vessel lifetime NDT values. (The fluences noted are maximum values found in Ref *1*. Projected fluences for a particular reactor vary even among tables of Ref *1*).

The surveillance programs outlined in Table 22 are remarkably uniform, representing largely Westinghouse systems. These usually include specimens from two or three of the core region plates. For the later plants eight capsules, containing in various arrays: Charpy V-notch, tension, and WOL specimens, representing the vessel plate ($1/4T$ location) plus the ASTM reference A302-B steel (a 6-in. plate). A more limited coverage of weld metal and HAZ also is provided. Capsules usually contain low melting point alloys for temperature monitoring and thermal and fast neutron monitors (iron, copper, and cobalt with and without cadmium shields). These capsules also contain neutron monitors of nickel, ^{238}uranium, and ^{237}neptunium. The schedule for capsule removal is usually after 1 1/2, 3, 5, 10, and 20 years with three capsules available for removal at other times deemed necessary.

The General Electric (GE) built reactors, which make up most of those listed in Table 23, also usually contain specimens from two to three of the core region plates. Twenty-three capsules usually are used, ten containing plate, weld, and HAZ Charpy-V specimens (one of these contains only a GE reference steel) and thirteen containing tension specimens. All capsules contain temperature and neutron monitors and are removed on a schedule of 2, 8, and 32 years for those plants where the vessel lifetime exposure approximates 1×10^{19}. If the fluence is much lower, as indicated (Table 23) for a number of plants, removal is on the basis of projected need.

Both the GE and Westinghouse programs follow rather closely the ASTM, Recommended Practice for Surveillance Tests on Structural Materials in Nuclear Reactors E 185 [*2*]. Unfortunately, the early reactor surveillance programs did not meet these high standards though, in retrospect, these reactors had a greater need for surveillance because of the projected higher fluences and related high projected NDT values.

It is unfortunate that of the 90 plus reactors which were operating, under construction, or planned by early 1969, data for only the 30 listed in Tables 22 and 23 are available thus far. Of course, the long lead for construction explains why so few firm vessel surveillance plans are yet available. It should be noted also that a positive response of providing future surveillance information to the MPC has been received from many

TABLE 24—*Chemical composition and heat treatment for Big Rock Point reactor surveillance base metal and weld metal composition* (*Ref* 4).

Material Analysis and Supplier	Chemical Analysis, weight %									
	C	Mn	Si	P	S	Ni	Cr	Mo	V	Cu
Base (GE)	0.18	1.43	0.25	0.016	0.021	...	...	0.51	...	...
Base (NRL)	0.30	1.42	0.25	0.016	0.018	0.18	0.13	0.51	0.02	0.10
Weld (NRL)	0.12	1.25	0.28	0.014	0.012	0.10	0.19	0.53	0.03	0.27

Heat Treatment *V*

Base metal: austenitized at 1600 ± 25 F for 4 h, brine quench; tempered at 1225 ± 25 F for 4 h, furnace cooled; stress relieved at 1125± 25 F for 20 h, furnace cooled.

Weld metal: preheat, interpass temperature, and postweld heat-treat cycles identical to those for lower shell section of the vessel.

of those who will own operating reactors from the 90 plus in some stage of development. These programs will be reviewed for content as part of a later effort of the Subcommittee on Effects of Radiation on Structural Metals.

The following reviews cover available data from several of the older reactors. In each case projections of fluence and NDT increase values to the end of planned life are made. In the critique of surveillance analysis results which then follows (Chapter 6) correlation between the real (observed) and the earlier calculated data will be made along with a brief review of the strong and weak aspects of each of the programs for which sufficient data are available to permit a meaningful analysis. The order of reactors covered by these reviews is approximate to the level of available information on the particular vessel surveillance program.

Big Rock Point Reactor

Surveillance Program

The Big Rock Point reactor (BRPR) is a 240 MWt boiling water reactor located near Charlevoix, Mich. The 6-in.-thick pressure vessel is fabricated of A302-B steel, the chemical composition and heat treatment of which are given in Table 24. The comprehensive surveillance program for this reactor has been described [*3*]. The program utilizes Charpy V-notch and tension specimens of BRPR base plate plus weld metal and heat affected zone (HAZ) of a fabrication test weldment duplicating the vessel construction practice. Included also is a standard General Electric reference material fabricated from the Humboldt Bay reactor vessel steel. The capsules also contained threshold flux monitors, low melting point alloy temperature monitors, and several sets of other neutron detector materials which may prove to be of value in the future.

The capsule irradiation locations are indicated schematically in Fig. 64. They include vessel wall locations, accelerated positions between the core and thermal shield, and above-core thermal control locations. The individual capsule loadings are described in Ref *4*. Several capsules containing most or all of the specimen materials are located in larger baskets at these locations. Withdrawal of a basket at a given time yields capsules with specimens required to make a complete periodic material evaluation for a specific operating interval.

Notch Ductility Properties

Three separate withdrawals of specimens from the BRPR have been made. At this time all of the Charpy-V specimens from the first two groups have been evaluated at NRL. These are evaluated on the basis of the 30-ft·lb level on the DBTT curve. The initial group of capsules provided specimens of the base metal, weld metal, and HAZ from both a vessel wall

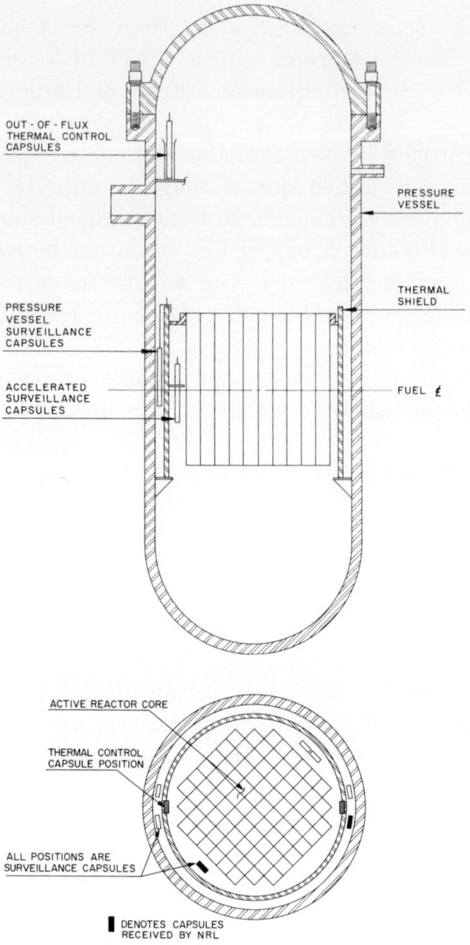

FIG. 64—*Schematic view of the Big Rock Point nuclear reactor, showing the relative locations of the surveillance capsule positions.*

location and an accelerated location. The results and analysis of these specimens have been reported [*5*]. The second withdrawal contained the same specimen types but were from a second accelerated location only. These data also have been reported [*6*]. The third withdrawal involved specimens of all types from both accelerated and vessel wall locations and have not been evaluated at this time.

Base Metal—The Charpy V-notch specimens were taken from the quarter thickness and were notched parallel to the plate surface. The results of evaluating the base metal specimens from the first two capsule withdrawals of the BRPR surveillance program are shown in Figs. 65 [*5*] and 66 [*6*]. In Fig. 65 it should be noted that the data points for the vessel

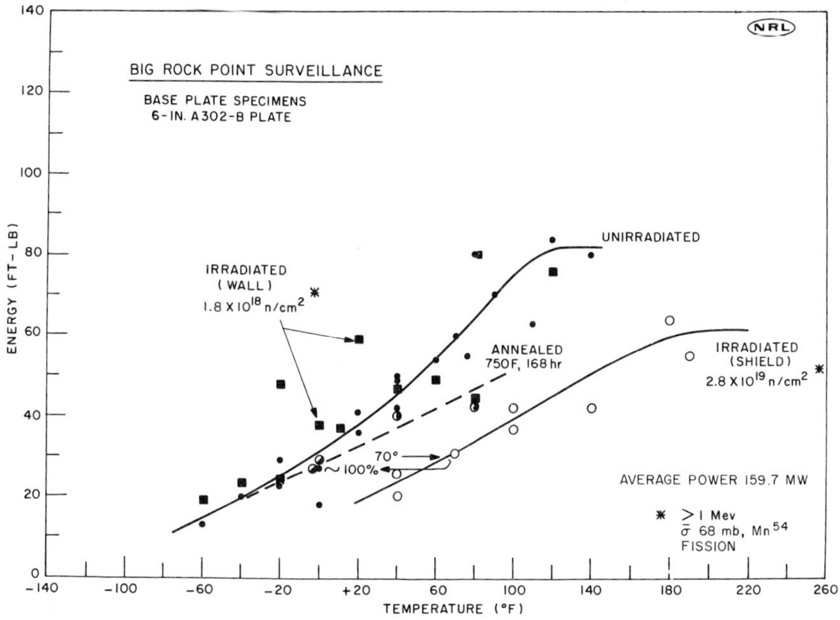

FIG. 65—*Charpy V-notch ductility characteristics of the Big Rock Point pressure vessel steel (6-in. A302-B base plate specimens) before and after irradiation. Data points for the vessel position (solid squares) fall within the scatterband of unirradiated data, while data points from the accelerated exposure, shield position (open circles) show a 70-deg Charpy V-notch 30-ft·lb transition temperature increase. Postirradiation annealing data (half-closed circles), although very limited, indicate almost full transition temperature recovery* (*Ref* 5).

wall location specimens, designated WALL, fall about the unirradiated condition curve indicating no measureable transition temperature increase. The base metal specimens irradiated at the accelerated location, designated SHIELD, did reveal a positively measureable increase for the fluence received. Evaluation of the second set (accelerated position) of base metal specimens, Fig. 66, revealed an additional increase for the significantly higher fluence. In both cases, a sufficient number of uniformly irradiated specimens were available for evaluation. Several of the specimens from the first accelerated group were given a postirradiation heat treatment of 750 F for 168 h. As shown in Fig. 65, virtually complete recovery was attained in terms of the transition temperature 30-ft·lb level, but the upper shelf energy absorption level was still somewhat degraded.

Heat Affected Zone—In the preparation of the HAZ specimens of the BRPR surveillance program, it was difficult to assure that the notch root would lie on the fusion line [*4*]. As a result, several of the HAZ specimens did not strictly represent the property they were intended to measure,

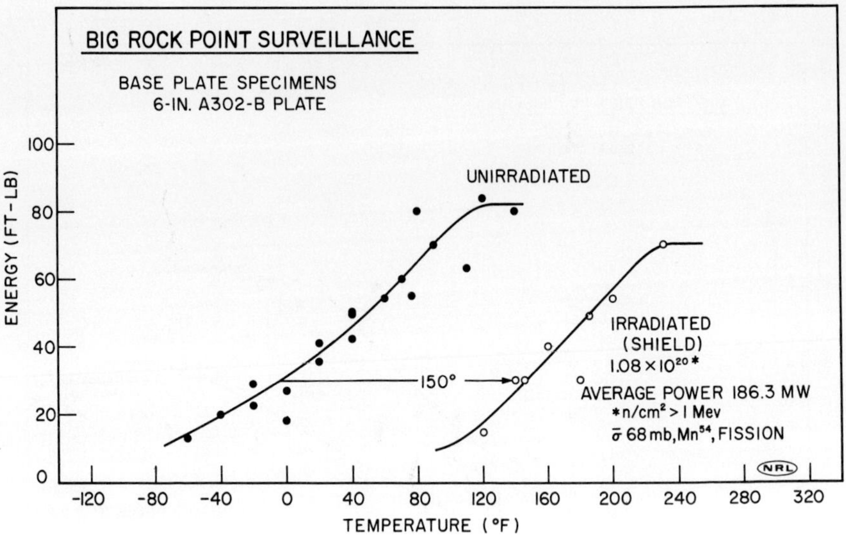

FIG. 66—*Charpy V-notch ductility characteristics of the Big Rock Point reactor pressure vessel steel (6-in. A302-B base plate) before and after irradiation in accelerated position No. 124 (Ref 6).*

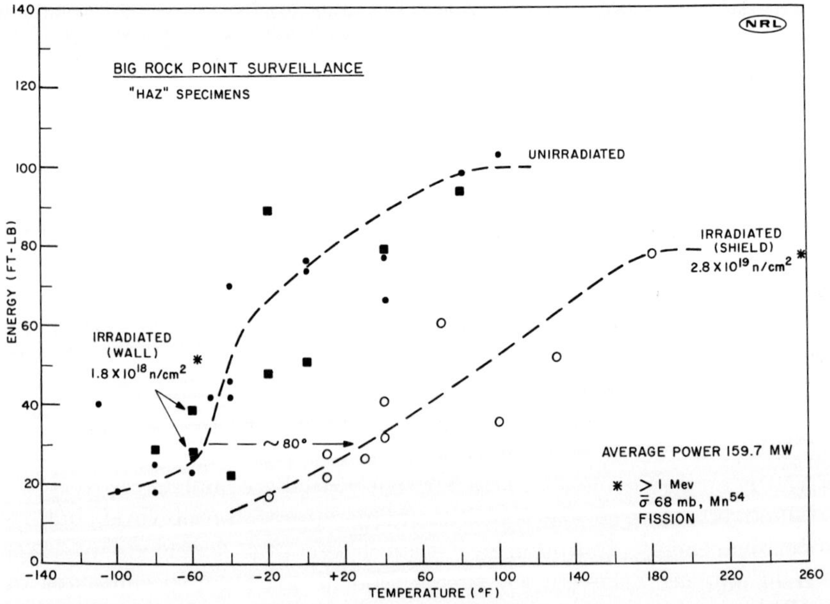

FIG. 67—*Charpy V-notch ductility characteristics of the Big Rock Point HAZ specimens before and after irradiation (Ref 5).*

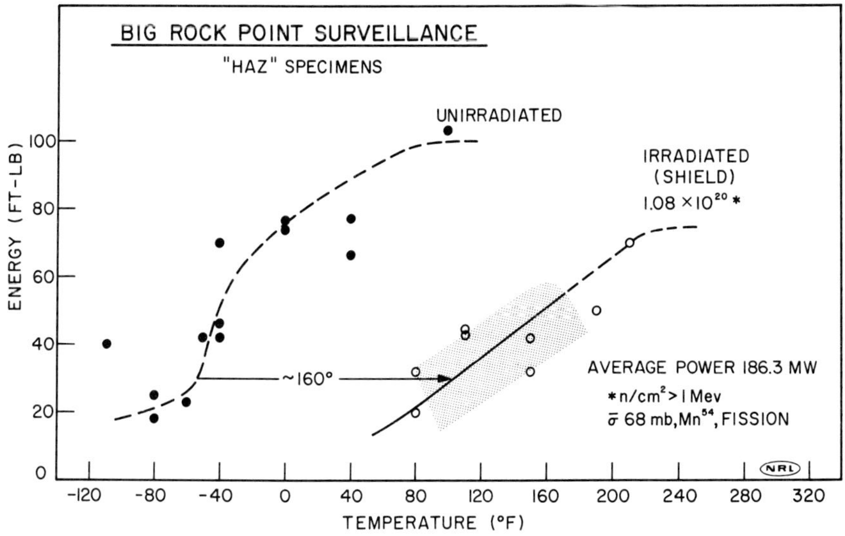

FIG. 68—*Charpy V-notch ductility characteristics of the Big Rock Point reactor HAZ specimens before and after irradiation in accelerated position No. 124. The band superimposed on the curve for the as-irradiated condition reflects the scatter of the data points* (*Ref* 6).

and thus yield a large degree of scatter. This scatter is evident from the results of the first withdrawal which are presented in Fig. 67 [*5*]. Here again, the data from the WALL location specimens fall within the scatterband of the unirradiated data; no measurable transition temperature increase is evident.

As with the base metal, a significant increase could be noted for the accelerated location (SHIELD) specimens although the scatter at higher energies presented a problem. Examination of the HAZ specimens from the second accelerated location withdrawal was quite difficult owing to the data scatter. As seen in Fig. 68 [*6*], it was necessary to depict the data in terms of a scatterband. Although the midpoint of the band was selected for assignment of transition temperature increase, the upper limit indicates a transition temperature of about +140 F which is consistent with the postirradiation transition temperature for the base metal (Fig. 66). Thus, the problems encountered in data interpretation for the HAZ specimens do not have significant consequences for analyzing the irradiated vessel condition.

Weld Metal—It is important to note that the weld metal (and HAZ) specimens were taken from a fabrication test plate which duplicated the practices employed on the actual vessel but did not come from an actual vessel weld, as from a nozzle dropout. This is true, of course, for most surveillance programs. Furthermore, the Charpy V-notch specimens were

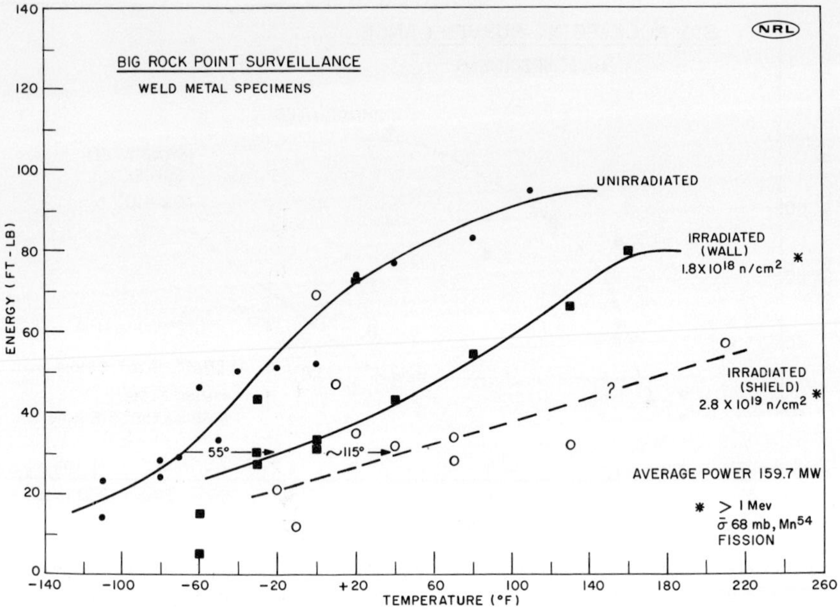

FIG. 69—*Charpy V-notch ductility characteristics of the Big Rock Point weld metal specimens before and after irradiation. Note positive Charpy V-notch 30-ft·lb transition temperature increase of both vessel wall and shield position specimens* (*Ref* 5).

machined with the notch root parallel to the plate surfaces, not perpendicular to the plate surfaces as is the usual practice. This orientation is considered acceptable if weld properties are uniform, but there is a general lack of knowledge as to such uniformity.

The specimens evaluated from the first withdrawal are shown in Fig. 69 [*5*]. In this case, the WALL location specimens clearly reveal a transition temperature increase for the low fluence received; this increase approaches the level reached by the base metal and HAZ specimens irradiated in the SHIELD location during the first surveillance period. The high degree of scatter for the SHIELD weld metal data is particularly evident in the temperature region between −20 and +20 F because of the trend of the data at higher temperatures; however, the assignment of a 30-ft·lb increase of about 115 F appears reasonable. Despite the significantly higher embrittlement increase for the weld metal, the 30-ft·lb transition temperature resides at about the same level as for the base metal and HAZ. Interestingly, and perhaps fortuitously, essentially the same situation applies for the analysis of the second withdrawal of accelerated specimens. The Charpy V-notch data shown in Fig. 70 [*6*] confirm the previous ob-

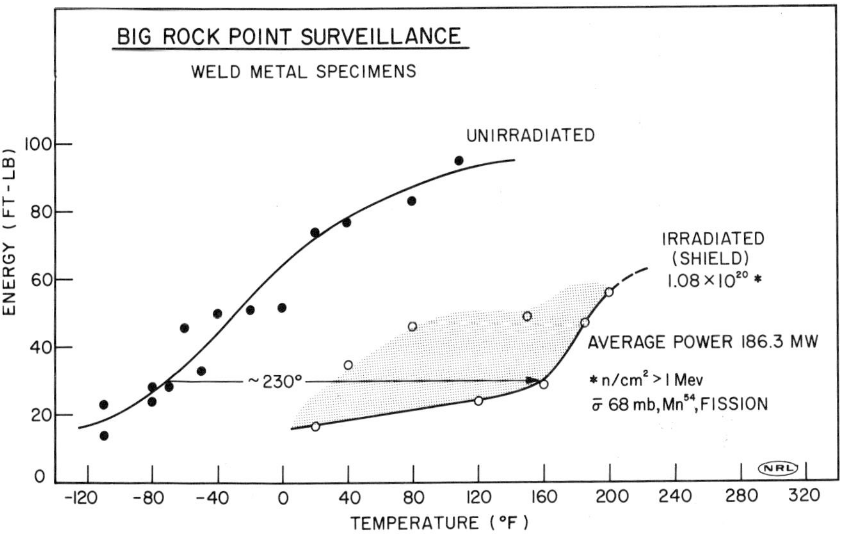

FIG. 70—*Charpy V-notch ductility characteristics of the Big Rock Point reactor weld metal before and after irradiation in an accelerated position No. 124. The placement of the specimen notch parallel rather than perpendicular to the weldment surface may have contributed to the degree of data scatter shown* (*Ref* 6).

servation of a more rapid weld metal transition temperature increase than shown by the base metal and HAZ. Again, the data scatter for the as-irradiated condition is quite severe. It can be speculated that the placement of the V-notches parallel rather than perpendicular to the plate surface may have contributed to the scatter. From the distribution of data points in Fig. 70, it was considered realistic to assign an upper limit (230 F) transition temperature increase [*6*]. In order to validate these results as being truly representative of the weld, three weld metal specimens broken at temperatures of +120, +160, and +185 F were polished and etched with a 1 percent nital solution to assure that the V-notches of the specimens were within the weld metal deposit [*6*]; this clearly was established. The low initial 30-ft·lb transition temperature of the weld metal, −70 F, places the irradiated-condition transition temperature at +160 F, or practically at the same as that of the base metal and of the upper limit of the HAZ band.

The limited data suggest that the upper energy shelf for the weld metal equals or exceeds 60 ft·lb, although the exact energy level could not be determined because of the small number of specimens available. The notch ductility data from the two surveillance specimen withdrawals of the BRPR are summarized in Table 25.

TABLE 25—*Summary of Big Rock Point reactor surveillance notch ductility results (irradiation temperature, ~585 F).*

Location and Material	Fluence[a] n/cm² > 1 MeV (fission)	Fluence[b] n/cm² > 0.5 MeV (calculated)	30-ft·lb Transition Temperature Increase, deg F			Energy Absorption at Full Shear, ft·lb	
			Unirradiated	Irradiated	ΔT	Unirradiated	Irradiated
Shield:							
Base	2.76×10^{19}	2.3×10^{19}	−5	65	70	82	62
Weld	2.76	2.3	−70	45	115	95	57
HAZ	2.76	2.3	−55	25	80	100	78
Shield:							
Base	10.8	9.0	−5	145	150	82	70
Weld	10.8	9.0	−70	~160	~230	95	> 56
HAZ	10.8	9.0	−55	~105	~160	100	> 70
Wall:							
Base	0.18	0.14	−5	−5	0	82	~ 82
Weld	0.18	0.14	−70	−15	55	95	80
HAZ	0.18	0.14	−55	−55	0	100	~100

[a] Fission spectrum σ = 68 millibarns (mb), ^{54}Fe (*n*, *p*) ^{54}Mn.
[b] Calculated spectrum.

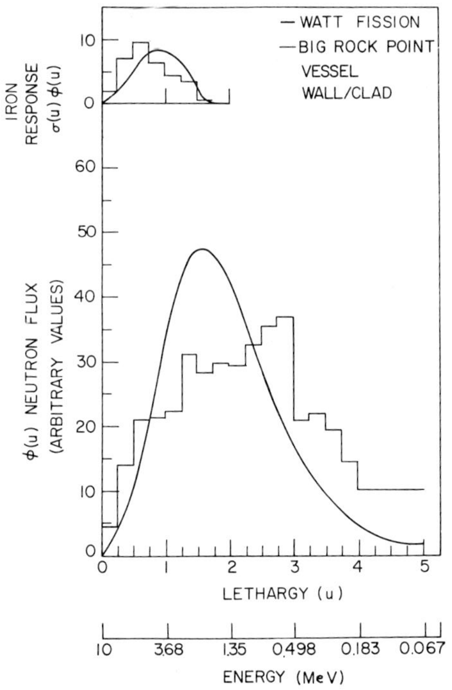

FIG. 71—*Graphical representation of the reactor energy distribution of the A 302-B/stainless steel clad interface of the Big Rock Point reactor versus the Watt fission spectrum plotted by equating the iron activation response in both spectra.*

Vessel Lifetime Projections

Neutron Dosimetry and Spectrum Considerations—In the foregoing BRPR analysis, the neutron fluences have been given as based upon the assumption of a fission spectrum distribution of neutrons at the various irradiation locations. Calculations of the theoretical neutron spectra for typical WALL and SHIELD locations have been obtained by NRL and are shown in relation to the Watt form of the fission spectrum in Figs 71 and 72, respectively. It is apparent that determination of neutron fluences, n/cm^2 >1 MeV, for the BRPR using the fission spectrum for the distribution of neutrons by energies does not adequately account for all neutrons. The main curves in both figures are keyed to the smaller curves in the upper portion of each figure which represent the activation of an iron neutron flux monitor wire. These curves are the product of flux times activation cross section for neutrons of the specific energies noted; they are quite abbreviated since no significant activation by the reaction, ^{54}Fe (n,p) ^{54}Mn, occurs in iron below about 1.5 MeV. Accordingly, there is essentially no response of iron to neutron flux below this point. Only the

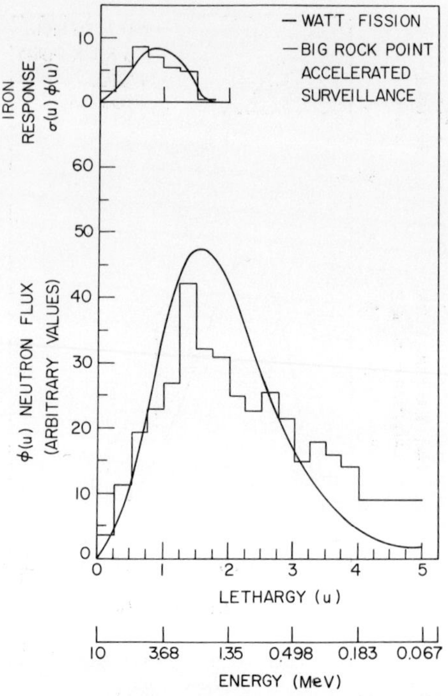

FIG. 72—*Graphical representation of the neutron energy distribution of the accelerated surveillance location of the Big Rock Point reactor versus the Watt fission spectrum plotted by equating the iron activation response in both spectra.*

neutrons above 1.5 MeV are effective in causing activation, so only those neutrons are represented in the upper portions of the figures. Development of the upper curves is detailed in Chapter 5. At this point it is only necessary to state that the smooth curve represents the iron activation if a fission spectrum were present, and the histogram represents the iron activation from the calculated spectrum. The upper curve relationships then dictate the relative magnitude of the calculated reactor spectra to the fission spectrum shown in the lower portion of each figure. From a given activation of an iron flux monitor wire in the BRPR, it is possible to calculate either a fission-spectrum-based fluence or a calculated-spectrum-based fluence, and the resulting values would be numerical images of the pictorial relationships in the figures.

The significance of the calculated spectrum is that it clearly reveals the BRPR spectrum to contain fewer neutrons of energies >1 MeV or >0.5 MeV than one would have projected if the fission spectrum had been the correct neutron distribution.

Projections—The neutron fluences for the BRPR surveillance data have been adjusted to account for the calculated-spectrum neutrons of energies

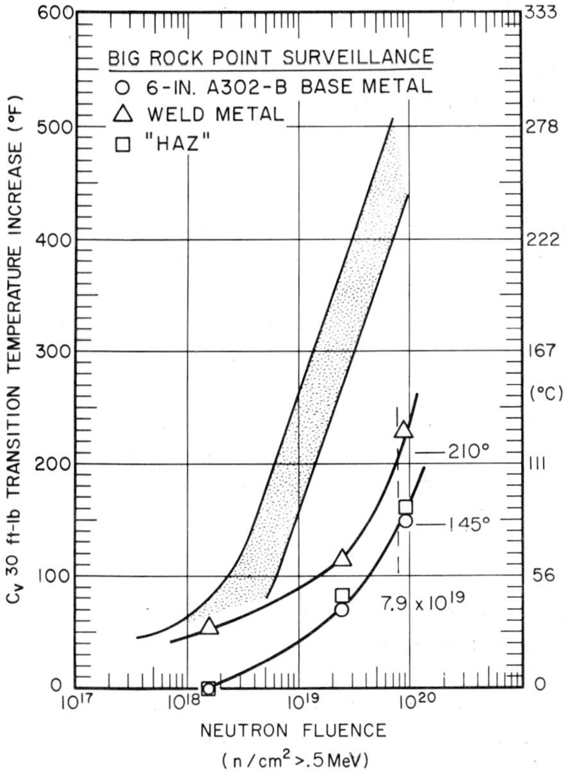

FIG. 73—*Increase in 30-ft·lb transition temperature versus calculated spectrum fluences >0.5 MeV for Big Rock Point reactor surveillance exposures. The 40-year service life is projected to yield a fluence of 7.9 × 10^{19} n/cm² >0.5 MeV.*

>1 MeV and >0.5 MeV in the respective irradiation locations. This extrapolation is discussed in more detail in Chapter 5. The fluences are shown in Table 25 along with the fission spectrum values for referral to the Charpy curves. The calculated-spectrum fluences are then used to plot the BRPR surveillance data in Fig. 73.

In summary, the flux level at the pressure vessel wall/stainless steel clad interface was 6.3 × 10^{10} n/cm²-s >0.5 MeV, or 2.79 × 10^{10} n/cm² >1 MeV. Integration over forty years yields a fluence of 7.9 × 10^{19} n/cm² >0.5 MeV or 3.5 × 10^{19} n/cm² >1 MeV for the BRPR. These values can be shown to be the likely peak values by reference to Fig. 64. Note that the vessel wall capsule is located where the fuel is closest to the vessel wall. Thus, the capsules apparently were located purposely to obtain peak exposures and thereby realistically project the point of highest embrittlement. Nevertheless, the lifetime limits must be based upon the maximum beltline fluence, and this is *it* for the BRPR; it yields a ΔT of

210 F for the weld versus 145 F for plate specimens. Considering an initial 30-ft·lb value of −70 F (weld) and −5 F (plate), the absolute ΔT for both is 140 F, a level too low to generate serious concern unless operating parameters change significantly.

Summary

The 30-ft·lb transition temperature increases developed from the BRPR surveillance program are plotted in terms of calculated neutron spectrum fluences in Fig. 73. The triangles represent the weld metal data and as such describe the highest embrittlement increases; the base metal and HAZ have responses similar enough to justify construction of a single line for the two materials. The >0.5 MeV, 40-year service fluence of 7.9×10^{19} is noted on the figure, and corresponding transition temperature increases for weld, base, and HAZ are indicated. The corresponding fluence >1 MeV is 3.5×10^{19}. It would appear from Fig. 73 that the weld metal increase trend will be the limiting condition for the BRPR pressure vessel over its lifetime. Since this figure only shows the increases, it is important to refer to Table 25 which gives the *initial* transition temperature for the three materials concerned. By referencing the weld metal increase to its initial transition temperature of −70 F, its as-irradiated 30-ft·lb transition temperature becomes +140 F. By interesting coincidence the base metal transition temperature becomes +140 F also (−5 F + 145 F = +140 F). Analogously, the 30-ft·lb transition temperature for the HAZ would be +90 F.

Aside from the fact that embrittlement cannot be foreseen as a problem to the BRPR, it is of interest to probe the potential causes for the variation in radiation response for the BRPR surveillance materials. A review of the chemical composition for the base and weld metals, shown in Table 24, is quite revealing. The wide difference between copper content in the base and weld metal can be taken as quite probably the prime reason for the differences in radiation response. It has been shown [*7*] that copper content in excess of 0.20 percent, the percentage contained in the ASTM reference A302-B steel, effects more embrittlement than that exhibited by the reference steel. Less copper than 0.20 percent works in the opposite direction. The weld metal data trend lies just to the right (less embrittlement) of the trend established for the ASTM reference A302-B steel after 550 F irradiation in test reactors [*8*]. This trend is to be expected in view of the about 585 F irradiation temperature of the BRPR weld metal. In fact, the very similar trends for the A302-B reference steel (550 F) and the BRPR weld (~585 F) suggest slightly greater sensitivity of the weld. Possibly because of the lower copper content of the base metal, its trend line falls considerably to the right of the weld metal and that for the ASTM reference A302-B steel. Again, the higher BRPR irradiation temperature

TABLE 26—*Chemical composition and heat treatment for 8⅛-in.-thick A302–B Yankee surveillance steel—Lukens Steel Melt No. 19281(Mark 8) (Ref 8).*

Evaluator	Chemical Analysis, weight %									
	C	Mn	P	S	Si	Ni	Cr	Mo	Al	Cu
Lukens	0.20	1.27	0.020	0.028	0.21	...	...	0.48	...	...
NRL	0.23	1.16	0.012	0.025	0.22	0.18	0.08	0.50	0.04	0.20

Heat Treatment

The heat treatment schedule followed by Babcock and Wilcox on the surveillance steel simulated the heat treatment given to the actual vessel-shell course plate. Heat treatment: austenitized at 1750 to 1800 F, spray quenched to below 600 F; stress relieved eight times at 1100 to 1150 F; soaking times based upon 1 h/in. of thickness. The as-fabricated vessel thickness was reduced from 8⅛ to 7⅞ in.

tends to push the trend to the right though there is no doubt about the lower embrittlement of the BRPR plate.

The very favorable service-life projection made for the BRPR should be tempered by the fact that both the base and weld metals did not receive all or precisely the same mechanical working and heat treatment as did the vessel. Nevertheless, the analysis of the BRPR was made relatively simple, yet virtually foolproof, because materials as fully representative as reasonably possible were placed at meaningful locations in the vessel for in-service irradiation. Regardless of how the radiation responses of the BRPR materials compared to that of other materials, positive behavior trends from the volume of irradiated specimens data were constructed easily for use in evaluating (and projecting into the future) the condition of the BRPR.

Yankee-Rowe Reactor

Surveillance Program

The Yankee reactor, a pressurized, light water plant of 600 MWt power located in Rowe, Mass., was constructed for the Yankee Atomic Electric Co. The 7 7/8-in.-thick pressure vessel is fabricated of ASTM A302-B steel; the chemistry and heat treatment are given in Table 26 [*8*]. The Yankee surveillance program steel was from a main vessel plate but was heat treated separately using the schedule given in Table 26.

The surveillance program for the Yankee vessel initiated at the beginning of Core II consisted [*9*] of ten elongated capsules each containing Charpy V-notch and tension specimens, arranged to provide eight at accelerated locations and two at locations adjacent to the vessel wall. In contrast to the BRPR surveillance program, the Yankee surveillance program provided for the inclusion of Charpy V-notch specimens of ASTM reference A302-B steel in each capsule. The information from these reference specimens was particularly useful in the overall analysis, since only limited capsules were available due to the loss of several capsules through failure of the capsule supports. No weld metal or HAZ specimens were included in the Yankee surveillance program.

A schematic view of the Yankee reactor (Fig. 74) locates the positions of the accelerated and vessel wall surveillance capsules. Two separate surveillance withdrawals were made. The first was after one single core life, Core II, when Capsules 1, 2, 6, and 8 were removed. After Core IV, the single remaining vessel wall capsule was removed.

Notch Ductility Properties

Accelerated Capsule Results—Evaluation of the CharpyV-notch results from the accelerated capsules was made difficult because of the large fluence differences between the capsules. It had been assumed that all four would have closely matching fluences due to their very similar vessel locations, but they apparently were displaced from these locations by water-flow turbulence.

The Yankee surveillance steel specimens from Capsule 8 were evaluated in the as-irradiated condition as were the reference steel specimens from all four capsules. The results for Charpy V-notch specimen evaluation

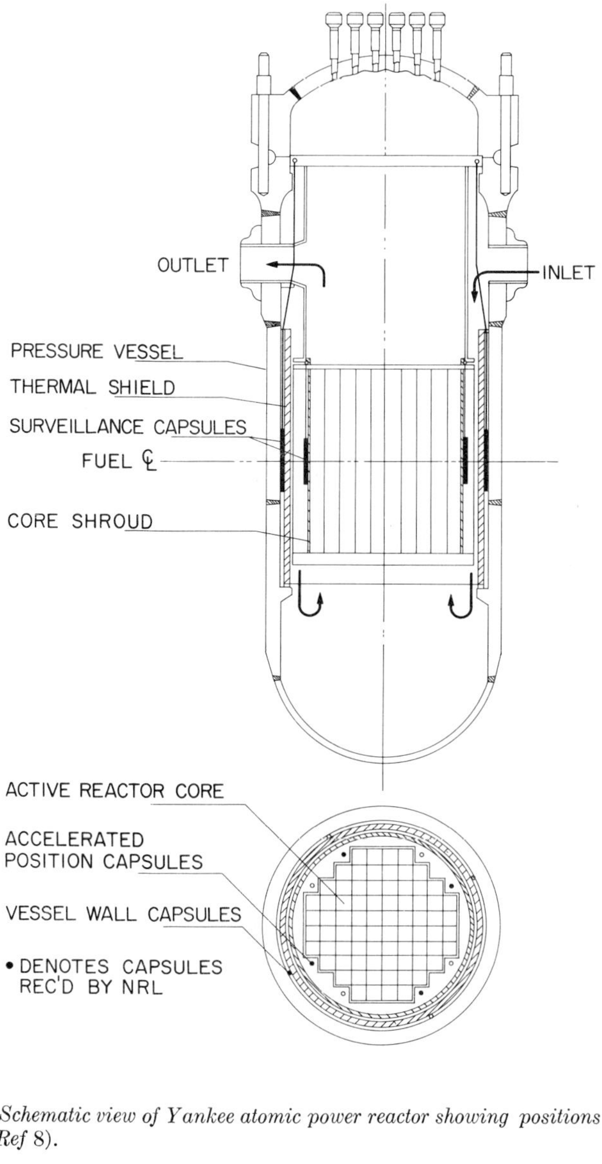

FIG. 74—*Schematic view of Yankee atomic power reactor showing positions and locations of capsules (Ref 8).*

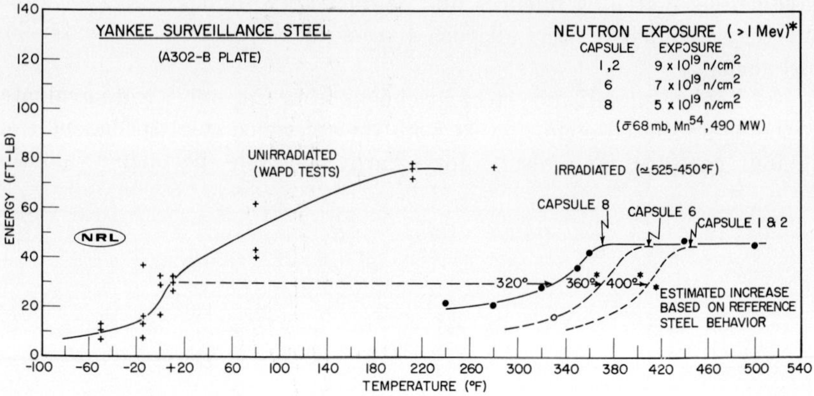

FIG. 75—*Charpy V-notch ductility characteristics of Yankee surveillance steel before irradiation and after exposure in accelerated surveillance locations in the Yankee reactor during Core II* (*Ref* 10).

of Yankee steel are given in Fig. 75 [*10*], which also shows estimated transition temperature increases for Yankee surveillance steel specimens of the other three capsules. One specimen from Capsule 6 was tested in the as-irradiated condition and used to help locate that particular curve. Limited data for the reference steel specimens (for which much experience has been developed), shown in Fig. 76 [*10*], reveal far smaller 30-ft·lb transition temperature increases than for the Yankee steel. The accelerated capsule surveillance results are summarized in Table 27 [*11*].

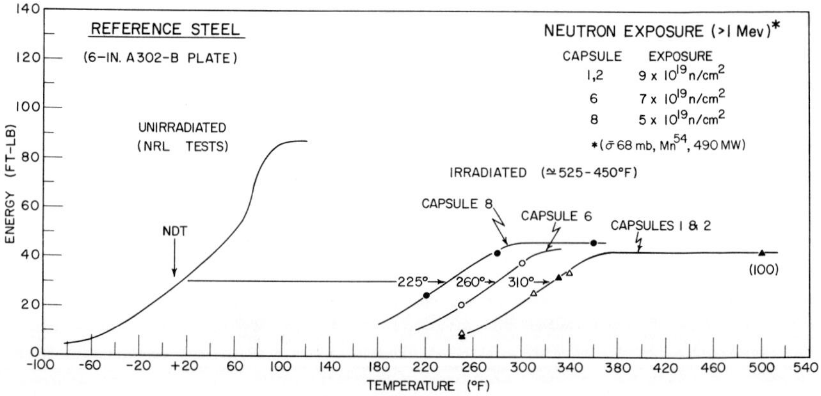

FIG. 76—*Charpy V-notch ductility characteristics of ASTM reference steel before irradiation and after exposure in accelerated surveillance locations in the Yankee reactor during Core II* (*Ref* 10).

TABLE 27—*Effect of irradiation on the notch ductility properties of Yankee surveillance specimens* (*Ref* 11).

Location, Capsule No., and Temperatures	Specimen Material	Neutron Fluence (n/cm²), fission spectrum		Charpy V-notch 30-ft·lb Transition Temperature, deg F			Energy Absorption at Full Shear, ft·lb		
		>1 MeV	2200 m/s	Unirradiated	Irradiated	ΔT	Unirradiated	Irradiated	Δft·lb
Vessel wall 520 to 450 F...	Yankee steel	2.2×10^{18}	...	10	120	~110	76	62	14
	Reference steel	2.2×10^{18}	...	35	120	85	87	62	25
Accelerated Capsule 1 525 to 450 F..........	Yankee steel	9×10^{19}	6.2×10^{20}	10	~410	~400	76	~45	~31
	Reference steel	9×10^{19}	6.2×10^{20}	15	325	310	87	42	45
Accelerated Capsule 2 525 to 450 F..........	Yankee steel	9×10^{19}	...	10	~410	~400	76	~45	~31
	Reference steel	9×10^{19}	...	15	325	310	87	42	45
Accelerated Capsule 6 525 to 450 F..........	Yankee steel	7×10^{19}	...	10	~370	~360	76	~45	~31
	Reference steel	7×10^{19}	...	15	275	260	87	44	43
Accelerated Capsule 8 525 to 450 F..........	Yankee steel	5×10^{19}	4.2×10^{20}	10	330	320	76	46	30
	Reference steel	5×10^{19}	4.2×10^{20}	15	240	225	87	46	41

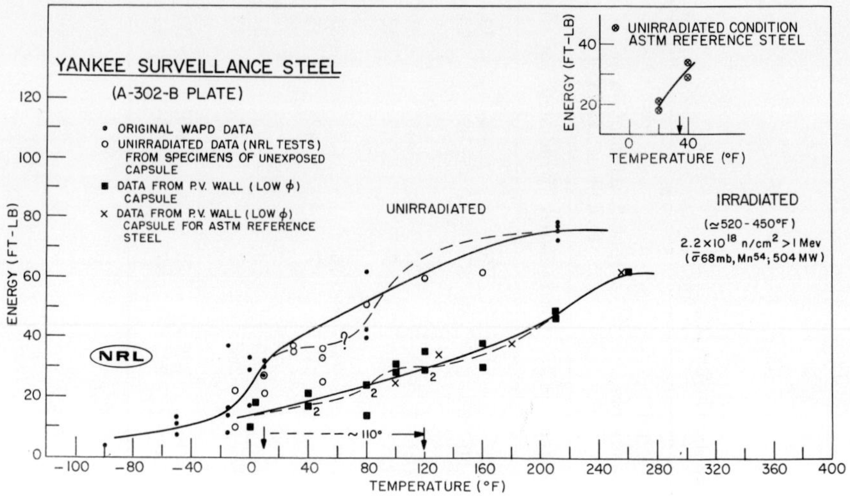

FIG. 77—*Charpy V-notch ductility characteristics of Yankee surveillance steel and ASTM reference steel before irradiation and after exposure at a vessel-wall surveillance location in the Yankee reactor during Cores II, III, and IV* (*Ref* 11).

Vessel Wall Capsule Results—The vessel wall results are of particular interest since they apply so directly to the temperature-spectral-flux conditions of the pressure vessel wall. The fluence for the single wall capsule was approximately one order of magnitude below the 30-cycle service fluence limit (projected from neutron dosimetry and spectral analysis data), despite the much longer exposure during Cores II, III, and IV. An extrapolation based upon this single capsule must be considered somewhat tenuous; confidence in the extrapolation, however, is gained from the data for the accelerated location.

Both the Yankee surveillance and the reference steel data for the vessel wall position are shown in Fig. 77 [*11*]. The unirradiated-condition solid curve is the same as that shown in Fig. 75. Additional unirradiated specimens of the Yankee surveillance steel were evaluated by NRL to provide data indicated by the open circles (the dotted line). Four additional unirradiated specimens of this reference steel were tested to define the 30-ft·lb transition temperature. These four data points are shown in the insert of the figure. They very clearly establish the initial 30-ft·lb value as +35 F. Turning to the irradiated-condition data, specimens of both the reference and Yankee surveillance steel fall about the same curve; the "x" points represent the reference steel and the solid squares represent the Yankee surveillance data. Both data curves indicate a 30-ft·lb transition temperature of +120 F which translate to an increase of 100 F for the Yankee surveillance steel and 85 F for the reference steel. The apparent double

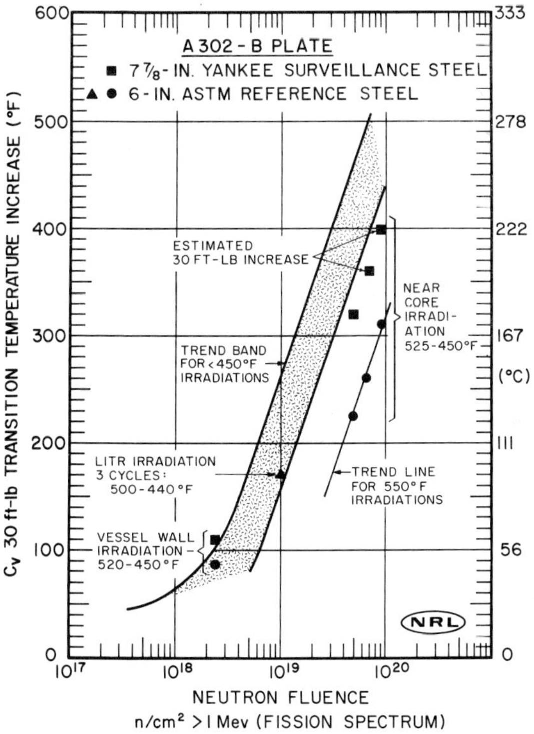

FIG. 78—*Increase in 30-ft·lb transition temperature versus fission spectrum neutron fluence >1 MeV for Yankee surveillance steel and ASTM A302-B reference steel for irradiations during Cores II, III, and IV* (*Ref* 11).

transition effect in the curve for the reference steel, shown as the dotted line and question mark, does not have any significant effect upon the increase. The vessel wall results are included in Table 27.

Discussion—The accelerated and vessel wall location data for both Yankee surveillance and reference steel have been plotted in Fig. 78 [*11*]. All the data presented are plotted versus neutron fluence based upon an assumed fission spectrum distribution of neutrons for all irradiation locations. A fission-spectrum-averaged cross section of 68 millibarns (mb) was used for the iron flux monitor reaction. The effect of neutron spectrum upon the Yankee reactor will be considered relative to vessel embrittlement projection here and in more detail in Chapter 5.

Inspection of Fig. 78 clearly shows the Yankee surveillance steel is more sensitive to neutron radiation than is the reference A302-B steel. The full reason for the great difference in sensitivity does not appear to lie with chemistry since both steels contained very similar amounts of the elements which have been identified as affecting embrittlement sensitivity (phos-

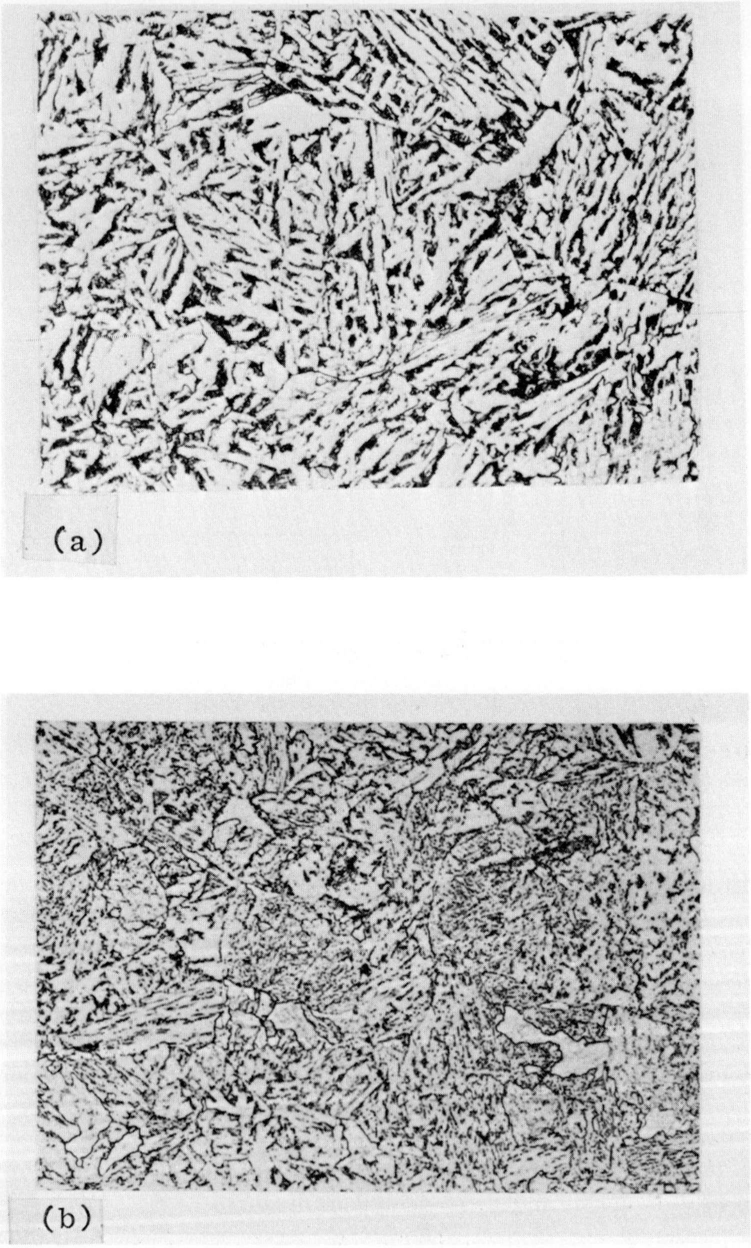

FIG. 79—*Microstructures of the Yankee surveillance steel and of the ASTM A302-B reference steel at ×500 (nital-picral etchant) (Ref* 11)*:* (a) *Yankee pressure-vessel surveillance steel; tempered upper bainite; and* (b) *ASTM A302-B reference steel; tempered upper bainite.*

phorous, copper, vanadium) [*7*]. The multiple stress relief cycles of the Yankee surveillance steel versus none for the reference steel might have created microstructure differences which may account for the variations observed, though the two have not been compared for this factor. However, a direct comparison of one stress relief cycle versus no stress relief on identical experimental laboratory-scale plates of A302-B steel showed no difference [*12*], but it must be pointed out that these experimental steels were quite "clean." The true role of residual element content in steels (relating to steel cleanliness) is not completely clear, but its importance appears to be growing. One additional factor should be noted. The microstructure of both the Yankee surveillance and reference steels were primarily tempered upper bainite (Fig. 79 [*11*], but the overall microstructure for the Yankee surveillance steel was quite coarse while that for the reference steel was relatively fine. A correlation between decreasing ferritic grain size and radiation sensitivity has been made [*13*], thus substantiating the possibility of microstructural effects.

A second interesting feature of Fig. 78 is that the vessel wall surveillance data of both steels appear to show more embrittlement than would be predicted on the basis of neutron fluence alone. The irradiation temperature, however, may have been responsible for the apparent better agreement of the accelerated data to the 550 F trend line than the vessel wall data. The most important fact is that the true irradiation temperature is not known and has only been estimated.

One significant fact of the temperature problem is that during the three core lives in consideration for the surveillance data, the vessel operating temperature declined steadily during the last third of each core cycle, as seen in Fig. 80. This temperature drop was from approximately 500 to 440 F. It had been shown previously [*14*] that increased radiation damage could occur for exposures at lower temperatures at the end of the irradiation period as in the Yankee reactor case. Research conducted specifically to duplicate the cyclic nature of the Yankee reactor vessel temperature, however, showed no difference in final transition temperature for ASTM reference steel A302-B specimens irradiated both at constant 500 F temperatures and to three complete cycles of two thirds at 500 F and a one third at steadily declining temperatures to 440 F as in the the reactor [*11*]. The short in-reactor life of the experiment compared to the three-year life of the vessel wall exposure period might have been effective in preventing some differentiation in data for the two conditions. The overall magnitude of the separation between Yankee surveillance and reference steel data from the Yankee reactor would appear to negate this argument.

There is presently no satisfactory reason that explains the wide sensitivity difference between the Yankee surveillance and the reference steels, nor is there an explanation for the higher embrittlement for both steels at the vessel wall location. The question of relating test reactor data with

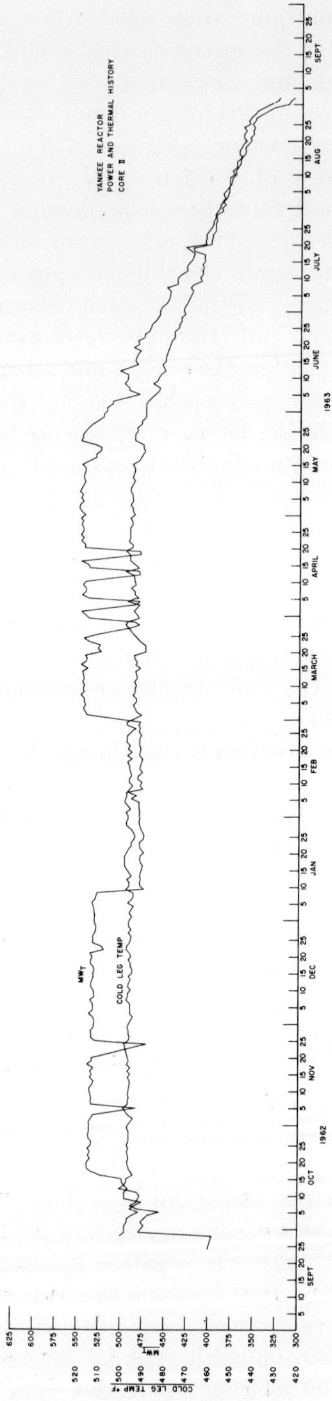
YANKEE REACTOR
POWER AND THERMAL HISTORY
CORE II
MWT
COLD LEG TEMP
COLD LEG TEMP °F
1962
1963

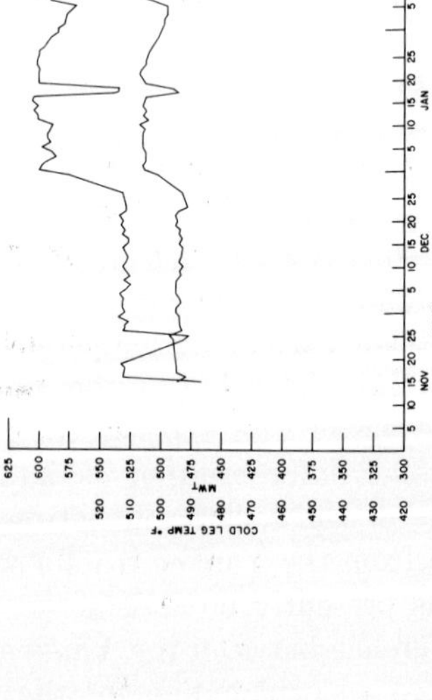
YANKEE REACTOR
POWER AND THERMAL HISTORY
CORE III
MWT
COLD LEG TEMP
COLD LEG TEMP °F
1963
1964

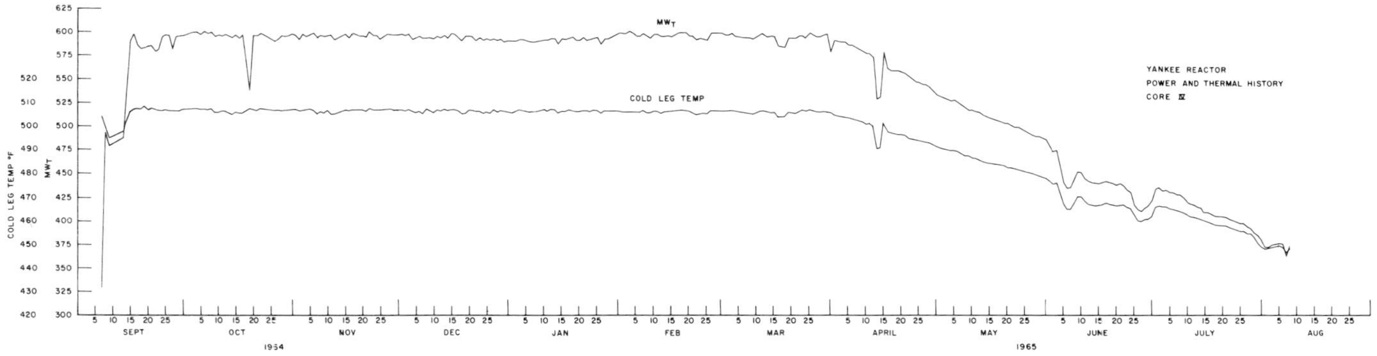

FIG. 80—*Inlet water temperature and thermal power levels for the Yankee reactor during Cores II, III, and IV (Ref 11).*

TABLE 28—*Effect of postirradiation annealing on the notch ductility properties of Yankee reactor surveillance steel specimens (approximated from limited data) (Ref 8).*

Capsule No.	Annealing Temperature, deg F	30-ft·lb Transition Temperature, deg F			Percent Recovery	Energy Absorption at Full Shear, ft·lb		
		Irradiated	Annealed	ΔT, deg F		Irradiated	Annealed	Δft·lb Recovery
2	850	410	10	400	100	45	76	31
1	750	410	120	290	72	45	64	19
6	640	370	210	160	45	45	53	8

power reactor data is of high importance and is presently under study for possible rate or spectrum effects, since there are several other cases wherein power reactor embrittlement data are more severe than would be predicted from test reactor data [*15, 16*].

Vessel Lifetime Projections

Two analyses have been made of the embrittlement/fluence conditions to be expected at the end of 30 full fuel cycles of the Yankee reactor at 600 MWt. The analysis by Westinghouse (WAPD) [*17*] yielded a value of 2.5×10^{19} n/cm^2 >1 MeV. It was developed using the P1MG diffusion code with additional correction factors for spatial, axial peaking, and corner effects. The results of this analysis predicted a 30-ft·lb transition temperature increase of 305 F. An important basis for this estimate was the neutron dosimetry performed to determine the flux and fluence values for the vessel wall capsule exposure. Seven of the irradiated Charpy specimens from the capsule of the Yankee surveillance steel were drilled out in the NRL hot cells with half of the drilling chips from each of

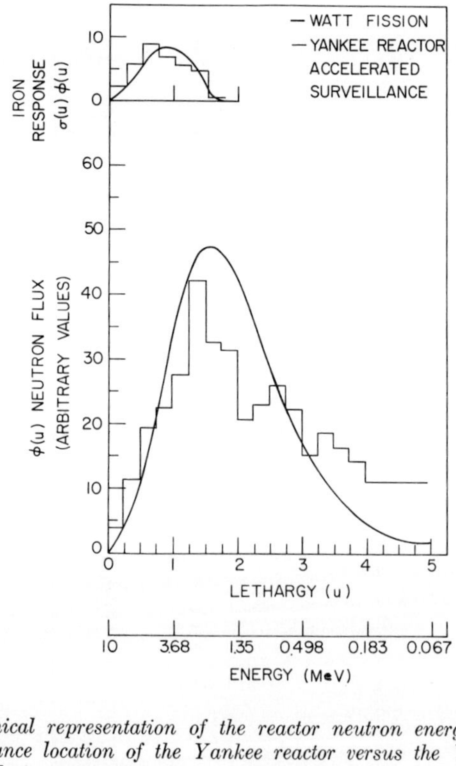

FIG. 81—*Graphical representation of the reactor neutron energy distribution of the accelerated surveillance location of the Yankee reactor versus the Watt fission spectrum plotted by equating the iron activation response in both spectra.*

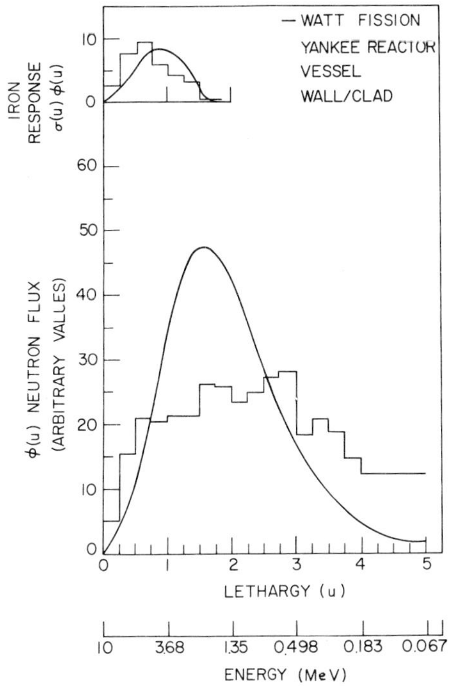

FIG. 82—*Graphical representation of the reactor neutron energy distribution of the A302-B/stainless steel clad interface of the Yankee reactor versus the Watt fission spectrun plotted by equating the iron activation response in both spectra.*

the seven specimens going to WAPD and the other half being retained by NRL for separate analysis. The results were in good agreement, but, upon extrapolation of the flux values over 30 fuel cycles of exposure time, some differences in the service-limit fluence could be expected.

The NRL analysis is described in Chapter 5 and will only be summarized here. Transport-theory computer analyses were performed to determine the Yankee reactor spectra. Those for the accelerated location and vessel wall adjacent to the stainless steel cladding are shown in Figs. 81 and 82 in relation to the Watt form of the fission spectrum. The NRL-developed 30-cycle service limit fluence determined from these spectral relationships and the dosimetry noted above was 1.46×10^{19} n/cm^2 >0.5 MeV, or 1.06×10^{19} n/cm^2 >1 MeV. Translation of the >0.5 MeV fluence into embrittlement required the plotting of the Yankee surveillance data in terms of calculated spectrum >0.5 MeV neutrons as shown in Fig. 83. This figure also shows 550 F transition temperature increase data for ASTM A302-B reference steel (irradiated in the low intensity test reactor), plotted in terms of the fission spectrum and the calculated spectrum >0.5 MeV neutrons [*11*]. A 265 F transition temperature increase was then indicated for the 1.46×10^{19} fluence by tracing along the left side of the 500 F WAPD reference curve at a distance defined by the solid square of the vessel wall Yankee surveillance.

The WAPD and NRL computer analyses of the Yankee reactor yielded relatively large differences in 30-cycle fluences. The initial flux value has been cited already as one reason for the difference, but a far more basic reason lies in the computer code libraries of nuclear data employed for the analyses. Neither library used the same values of microscopic cross sections for scattering, absorption, etc. for all of the many discrete energy groups. By this simple fact, then, the two spectral calculations and their results cannot be fairly compared. The data presented in Fig. 83, however, were all calculated with transport-theory codes which used the same cross section libraries, thus permitting meaningful comparisons.

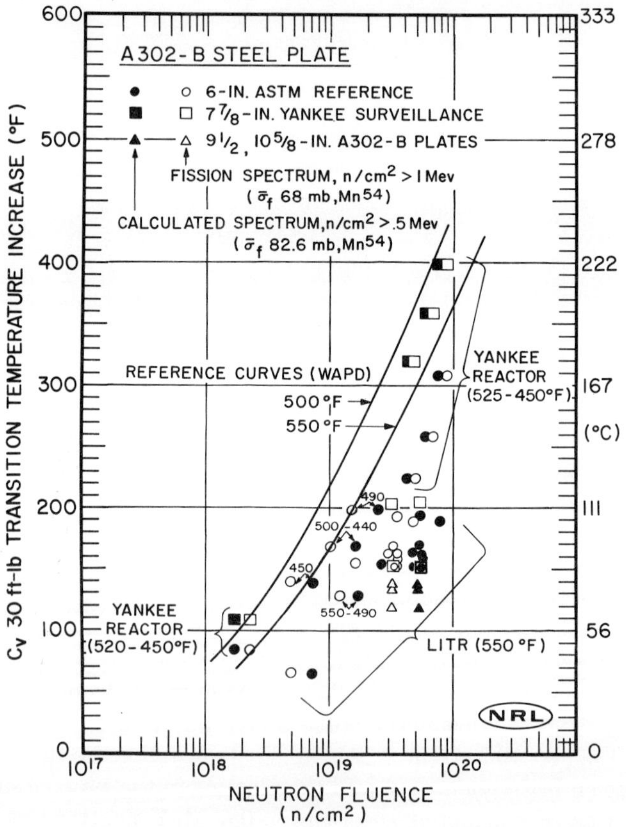

FIG. 83—*Transition temperature and NDT increase (indexed to Charpy V-notch 30-ft·lb level) versus neutron fluence determined using an assumed fission spectrum for interrelating the >1 MeV data (open points) and determined using calculated spectra for interrelating the >0.5 MeV data (closed points)* (*Ref* 11).

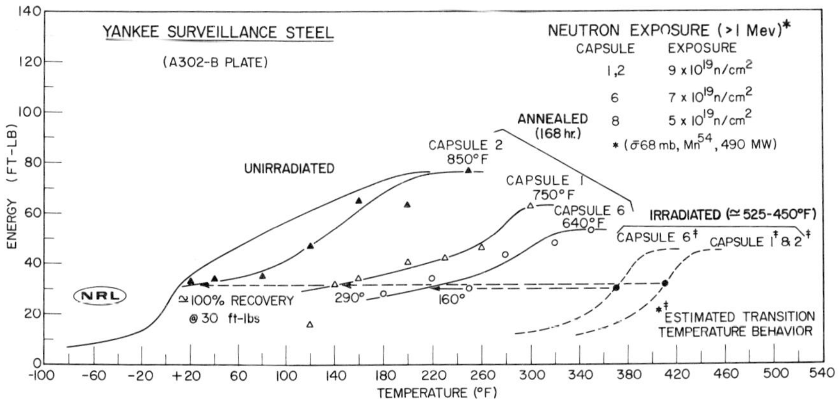

FIG. 84—*The effect of postirradiation heat treatment on the Charpy V-notch ductility characteristics of Yankee surveillance steel after irradiation in accelerated surveillance locations in the Yankee reactor during Core II* (*Ref* 8).

An important facet of both lifetime analyses is that they assume full power, full time operation for 30 years to reach the maximum transition temperature increases. Over a 30 calendar year period, the Yankee reactor most certainly cannot operate for a full 30 years at 600 MWt. Accordingly, the maximum vessel embrittlement necessarily will be less than the two ΔT values quoted (Westinghouse-305 F, NRL-265 F) unless the core configuration is changed (as it has been before) to again accentuate the beltline vessel flux or unless the power level is again raised.

Embrittlement Relief

Results of postirradiation annealing at three different temperatures (640, 750, and 850 F) for 168 h upon the Yankee surveillance steel specimens from Capsules 1, 2, and 6 of the initial surveillance withdrawal are presented in Fig. 84 [*8*]. The magnitude of notch ductility recovery for material contained in Capsules 1, 2, and 6 could not be determined with high accuracy, since the reference curves for the as-irradiated condition are "best estimates" based on annealing studies which were initiated before the assumption that all capsules receiving the same fluence had been tested. Test results for the postirradiation annealed specimens are summarized in Table 28 [*8*].

Through Thickness Embrittlement Evaluation

The Yankee reactor analysis thus far has been concerned with the A302-B steel at the interface of the alloy steel and stainless steel cladding. This location will likely sustain the highest embrittlement, with the embrittlement dropping off as the neutrons pass through the wall thickness.

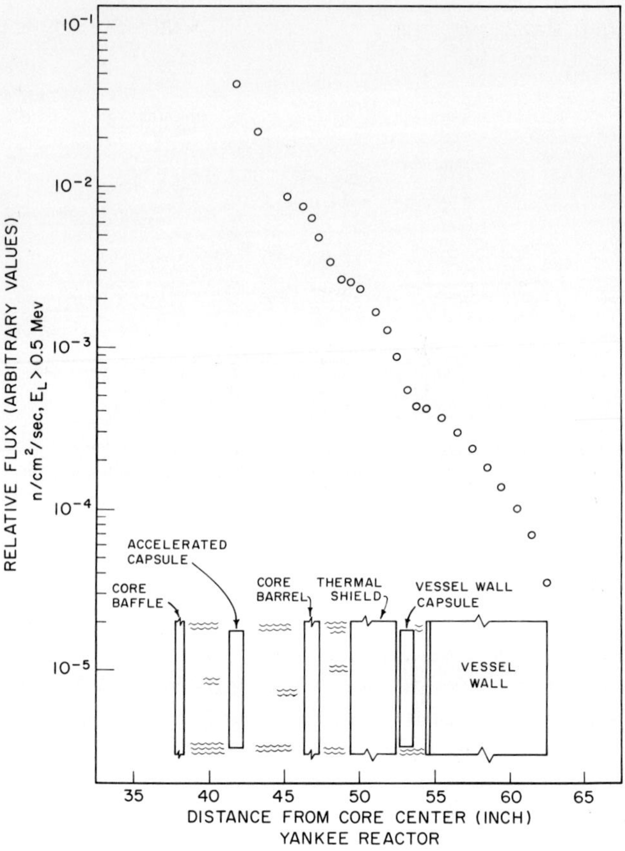

FIG. 85—*Graphical representation of the neutron flux intensity from the fuel core of the Yankee reactor to the exterior of the pressure-vessel wall. The neutron fluxes are derived from a calculated spectrum and only account for neutrons whose energies are greater than the lower energy limit* (E_L) *>0.5 MeV. The low values for the fluxes are simply peculiarities of the computer program output and are not to be construed as real flux values* (*Ref* 11).

The flux degradation across the Yankee reactor vessel wall was calculated and is presented schematically in Fig. 85 [*11*]. These values also are tabulated in Table 29 [*11*] with estimated transition for the A302-B vessel steel at these locations. Clearly, a distinct gradient in embrittlement exists in the vessel wall. For example, for a point midway through the vessel, the level of embrittlement has dropped more than 60 F. Realizing that the initial NDT at the center of thick plates is often higher, however, this factor must be minimized unless the vessel has been fully characterized. However, this computation demonstrates that a reserve of ductility can be expected in any reactor vessel relative to the condition at the alloy steel/stainless steel cladding interface. This, of course, depends upon the

TABLE 29—*Relative neutron fluxes, fluences, and corresponding 30-ft·lb ΔNDT values in the Yankee reactor reflector region and pressure-vessel wall* (*Ref* 11.)

Location	Distance from Core Center, in.	Relative Flux[a] >0.5 MeV	Φ_{cs} (>0.5 MeV) (n/cm^2)	ΔNDT, deg F
Accelerated capsule	41.85	4.27×10^{-2}	...	...
Water	43.35	2.17×10^{-2}	...	...
Water	45.35	8.67×10^{-3}	...	...
Core barrel	46.35	7.42×10^{-3}	...	...
Core barrel	46.85	6.39×10^{-3}	...	...
Core barrel	47.35	4.75×10^{-3}	...	...
Water	48.07	3.36×10^{-3}	...	...
Water	48.78	2.61×10^{-3}	...	...
Thermal shield	49.50	2.51×10^{-3}	...	...
Thermal shield	50.10	2.29×10^{-3}	...	...
Thermal shield	51.20	1.63×10^{-3}	...	...
Thermal shield	51.90	1.27×10^{-3}	...	...
Thermal shield	52.50	8.69×10^{-4}	...	...
Vessel-wall capsule	53.25	5.42×10^{-4}	...	...
Water	53.87	4.24×10^{-4}	...	...
Water/stainless steel clad interface	54.50	4.18×10^{-4}	...	...
0.1 in. in wall	54.61	3.98×10^{-4}	1.46×10^{19}	265
1 in. in wall	55.50	3.67×10^{-4}	1.34×10^{19}	255
2 in. in wall	56.50	2.96×10^{-4}	1.09×10^{19}	240
3 in. in wall	57.50	2.33×10^{-4}	0.85×10^{19}	210
4 in. in wall	58.50	1.80×10^{-4}	0.66×10^{19}	190
5 in. in wall	59.50	1.36×10^{-4}	0.50×10^{19}	170
6 in. in wall	60.50	9.90×10^{-5}	0.36×10^{19}	145
7 in. in wall	61.50	6.81×10^{-5}	0.25×10^{19}	125
8 in. in wall	62.50	3.47×10^{-5}	0.13×10^{19}	90

[a] The low values of fluxes are only peculiarities of the computer program output. They may be multiplied by some value such as about 10^{15} to conform to reality.

interface condition. The ability of this more ductile steel to resist brittle fracture should be evident from the PM-2A reactor vessel analysis.

Summary

Charpy V-notch specimens from the A302-B Yankee reactor surveillance steel irradiated in two surveillance locations of the Yankee reactor along with Charpy V-notch specimens of a reference A302-B steel have been tested and evaluated. The impact data show that the Yankee surveillance steel, from one main reactor vessel plate, but heat treated separately, is more sensitive to embrittlement by high energy neutron radiation than is the A302-B reference steel of the same nominal composition.

Theoretical neutron spectral calculations for the two surveillance locations in the Yankee reactor as well as for the interface between the alloy steel pressure vessel and stainless steel clad revealed significant deviations from a fission spectrum. By adjusting fluxes based on a fission spectrum at the surveillance locations to comply with calculated spectra for those locations, it has been possible to extrapolate these adjusted flux values from surveillance locations to the pressure vessel wall and to project a maximum neutron fluence for the Yankee reactor of 1.46×10^{19} n/cm^2 >0.5 MeV, or 1.06×10^{19} n/cm^2 >1 MeV.

The maximum neutron fluence calculated for the Yankee reactor in terms of n/cm^2 >0.5 MeV, derived from a calculated spectrum for the reactor, has been entered on a graphical representation of all the transition temperature increase data developed by NRL for elevated temperature irradiations of A302-B steel for which calculated spectrum fluences are available. A careful evaluation of the data [*11*] revealed that the maximum transition temperature increase for the Yankee reactor after 30 cycles of operations at 600 MWt will be about 265 F against a projected ΔT of 305 F by Westinghouse. Further investigations regarding the degradation of embrittlement and neutron fluence through the thickness of the Yankee reactor vessel wall also revealed that the neutron fluence level, based upon a calculated neutron spectrum, drops 76 percent in 6 in. across the thickness of the alloy steel vessel wall. Accordingly, the embrittlement level should also drop significantly across the thickness of the vessel wall.

The available data for the Yankee plant would seem to present a fair case for safe operations throughout the reactor lifetime. It must be recalled, however, that the heat treatment for the surveillance steel was not the exact treatment given to the main vessel plates; the surveillance steel only received a heat treatment to simulate that received by the main vessel plates. Perhaps an even more significant area of doubt is that of the weld metal and HAZ properties after irradiation, and indeed, even in the preservice condition. For example, the weld metal could have a higher copper content partially contributed by a flash coating of copper on the welding electrodes. And, as indicated in Chapter 2, when copper is higher in the weld than the base metal, greater embrittlement could accrue to the vessel. Having no knowledge of weld properties leaves the requirement of assuming the worst.

Doubt about the temperature of the surveillance specimens relative to the vessel wall is another area of question. Interpretation of data was hampered seriously by this unknown. An attempt to overcome this doubt could have been made using low melting alloy temperature monitors located within the capsules.

Although a great deal is known of the Yankee reactor through its limited surveillance program, certain details are unknown quantities because of the inadequacy of the overall surveillance program. These several questionable areas raise a degree of doubt regarding the embrittlement condition of the Yankee reactor which suggests need for care in changing core loading or other operational factors.

Dresden Nuclear Power Station

Surveillance Program

The initial reactor plant at the Dresden Nuclear Power Station (DNPS), Morris, Ill., is a 600 MWt boiling water reactor fabricated of 5 3/4-in.-thick A302-B steel. The chemical composition and heat treatment for both the pressure vessel and a plate section from a run-out of vessel course 6 adjacent to the fuel, used for the surveillance program material, are given in Table 30 [*18*].

The surveillance program initially designed for the reactor consisted of both longitudinal and transverse Charpy V-notch notch impact and tension specimens machined from surface thickness vessel plate material rather than 1/4T material; specimen designations were inside vessel surface) and outside. The notch root of the Charpy V-notch specimens was parallel rather than perpendicular (conventional practice) to the plate surface. After the initial installation, additional surveillance capsules were inserted containing Charpy and tension specimens of weld metal and weld HAZ material. It is assumed that these specimen materials were prepared the same as for the Big Rock Point program: weld Charpy specimens notched parallel to the plate surface with several specimens in which the notch root was not directly on the fusion line of the HAZ. Tension specimens of all materials had 1/4-in.-diameter gage sections.

TABLE 30—*Chemical composition and heat treatment for Dresden Nuclear Power Station reactor pressure vessel steel* (*Ref* 18).

Material Identification	Chemical Analysis, weight %					
	C	Mn	Si	P	S	Mo
Lukens Steel Co. Melt 19716, Slab 3B..................	0.21	1.33	0.20	0.018	0.018	0.45
Heat Treatment						
All plates fabricated and heat treated by Lukens Steel. Austenitized at 1650 to 1700 F, spray quenched to 1000 F in 10 min; tempered at 1200 to 1250 F, air cooled. The pressure vessel was postweld stress relieved at 1150 F periodically, and furnace cooled; the surveillance material was postweld stress relieved for a minimum of 29 h and a maximum of 50 h. During specimen fabrication the surveillance steel was stress relieved by heating to 1150 F in 4 h and holding at 1150 F for 1 h/in. of thickness followed by furnace cooling.						

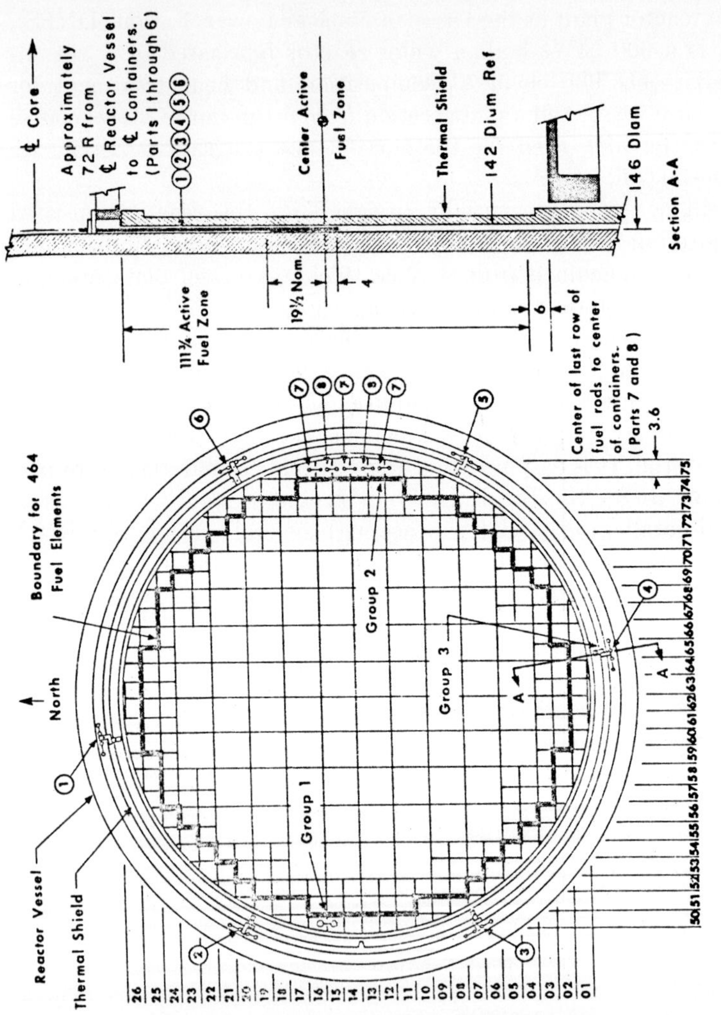

FIG. 86—*Schematic view of the DNPS reactor core showing relative location of the surveillance capsules (Ref 19).*

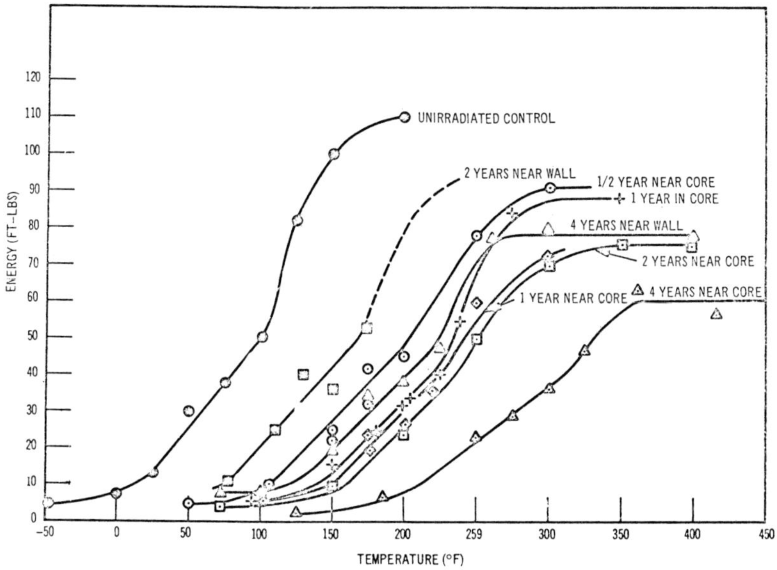

FIG. 87—*Charpy V-notch ductility properties of inside longitudinal* (o) *base material* (*Ref* 19).

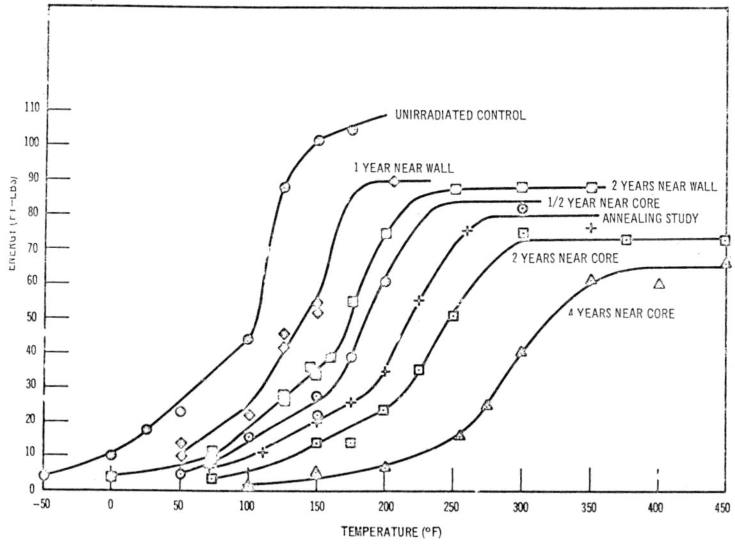

FIG. 88—*Charpy V-notch ductility properties of outside longitudinal* (ooo) *base material* (*Ref* 19).

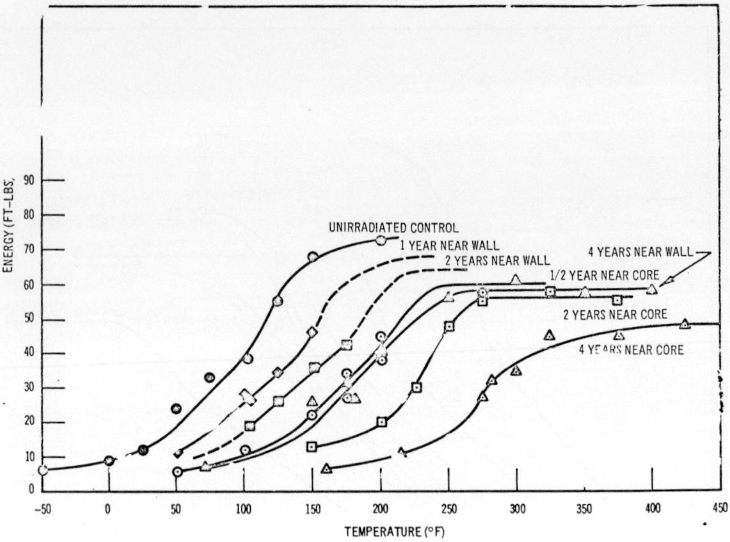

FIG. 89—*Charpy V-notch ductility properties of inside transverse* (oo) *base material* (*Ref* 19).

Provision was made for irradiation of these specimens at accelerated, vessel wall, and thermal control (designated turning vane or steam drum) locations [*18*]. The DNPS surveillance program, initial results, and subsequent surveillance data have been published [*19*]. The reactor core ar-

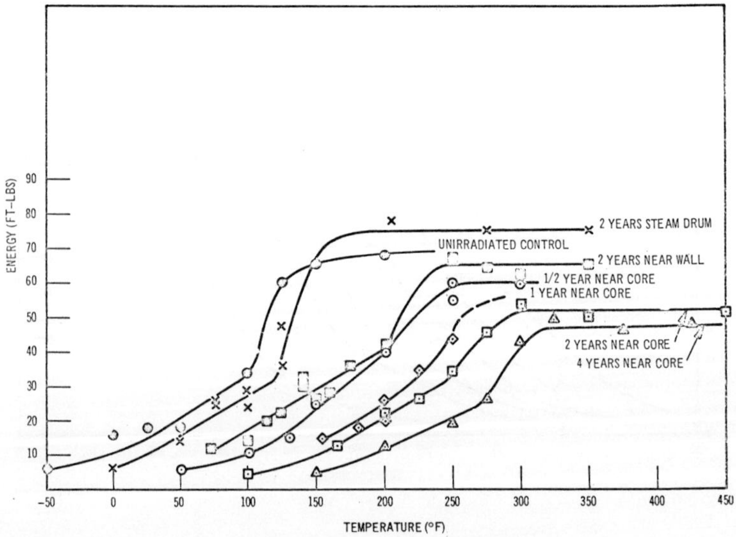

FIG. 90—*Charpy V-notch ductility properties of outside transverse* (oooo) *base metal* (*Ref* 19).

rangement plus a schematic view of some of the surveillance capsule locations is shown in Fig. 86.

Notch Ductility Properties

The Charpy V-notch properties of the inside and outside surface of the base metal, in both the longitudinal and transverse orientations before and after irradiation, are shown in Figs. 87 through 90, respectively. The Charpy V-notch properties of the surveillance weld metal and HAZ both before and after irradiation are shown in Figs. 91 and 92, respectively. The Charpy V-notch surveillance data are summarized in Table 31 and presented graphically in Fig. 93 [*20*].

The most interesting feature of the DNPS surveillance program is the exceptionally high initial 30-ft·lb transition temperature for steel taken from the plate surface layers. Specifically, the +60 and +90 F values for longitudinal and transverse specimens, respectively, should be of concern because specimens taken in this manner usually represent the best steel condition for the plate.[2]

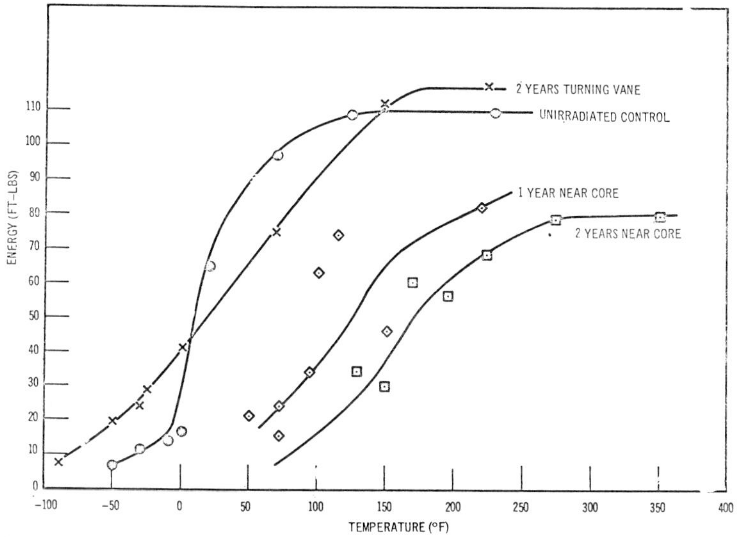

FIG. 91—*Charpy V-notch ductility properties of weld metal* (*Ref* 19).

[2] The preirradiation NDT characteristics of plate surface steel are generally superior to steel farther toward the plate center. This has been shown in Chapter 1 (Fig. 12) and in the previous section on the Yankee reactor. The differences may be great enough that the advantages gained by flux attenuation across vessel walls may be offset somewhat by the higher initial NDT values. The irradiated NDT at the quarter or half thickness of a heavy-sectioned plate could be higher than the NDT at the surface of the plate. Because the preirradiation NDT characteristics of the DNPS steel were not characterized through thickness, a conservative estimate of the quarter and half thickness NDT values is required. Accordingly, the surveillance projections for the DNPS must be conservative in line with these facts.

TABLE 31—*Average Charpy V-notch 30-ft·lb transition temperatures and shifts* (*Ref* 19).

Material and Code	Orientation[a]	Irradiation Exposure			Equivalent Years at Wall	Integrated Neutron Flux, nvt > 1 MeV		30-ft·lb Transition Temperature, deg F		30-ft·lb Transition Temperature Shift, deg F
		Group	Location	Years		Calculated	Measured	[c]	[d]	
Base o, ooo	L	...	control	...	0	...	...	60	60	...
		3	near wall	1	1.6	0.5×10^{18}	0.5×10^{18}	...	105	45
		4	near wall	2	3.2	1.4×10^{18}	4.7×10^{18}	125	135	75
		5	near wall	4	6.4	2.9×10^{18}	9.8×10^{18}	180	...	120
		6	near core	½	25	0.5×10^{19}	0.7×10^{19}	155	160	100
		7	near core	1	50	1.5×10^{19}	1.5×10^{19}	200	...	140
		8	near core	2	100	3.4×10^{19}	4.8×10^{19}	215	215	155
		9	near core	4	200	8.4×10^{19}	17.5×10^{19}	280	285	225
		12	in core	1	~ 75	...	6.1×10^{19}	190	...	130
Base o, ooo	L[b]	10	near core	1 + 1	~150	...	1.3×10^{20}	...	190	130
Base oo, oooo	T	...	control	...	0	...	...	80	90	...
		1	steam drum	2	0	$\sim 3 \times 10^{12}$	...	...	110	25
		3	inside wall	1	1.6	0.5×10^{18}	0.5×10^{18}	115	...	30
		4	inside wall	2	3.2	1.4×10^{18}	4.7×10^{18}	135	155	60
		5	inside wall	4	6.4	2.9×10^{18}	9.8×10^{18}	180	...	95
		6	near core	M	25	0.5×10^{19}	0.7×10^{19}	170	170	85
		7	near core	1	50	1.5×10^{19}	1.5×10^{19}	...	215	130
		8	near core	2	100	3.4×10^{19}	4.8×10^{19}	225	235	145
Base oo, oooo	T	9	near core	4	200	8.4×10^{19}	17.5×10^{19}	280	280	195

Standard oo	T	...	control	...	0	...	...	10	...
(Humboldt) oo	↕	2	turning vane	2	0	~3 × 10^{14}	<2 × 10^{14}	10	0
↓ ↓		10	near core	1	~150	...	1.3 × 10^{20}	165	155
	T	11	near core	2	~250	...	2.3 × 10^{20}	220	210
Weld DW	...	...	control	...	0	...	...	− 0	...
↓ ↓		2	turning valve	2	0	~3 × 10^{14}	> 2 × 10^{14}	−20	−20
		10	near core	1	~150	...	1.3 × 10^{20}	85	85
		11	near core	2	~250	...	2.3 × 10^{20}	135	135
HAZ DH	...	...	control	...	0	...	...	−25	...
↓ ↓		2	turning vane	2	0	~3 × 10^{14}	>2 × 10^{14}	−55	−30
		10	near core	1	~150	...	1.3 × 10^{20}	120	145
		11	near core	2	~250	...	2.3 × 10^{20}	200	225

[a] L = longitudinal, T = transverse.
[b] Annealing study.
[c] Inside surface.
[d] Outside surface.

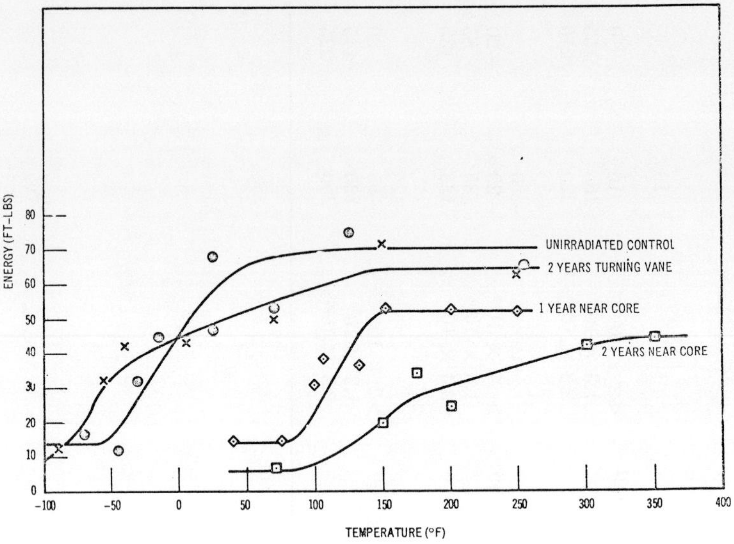

FIG. 92—*Charpy V-notch ductility properties of HAZ material (adjacent to the weld) (Ref 19).*

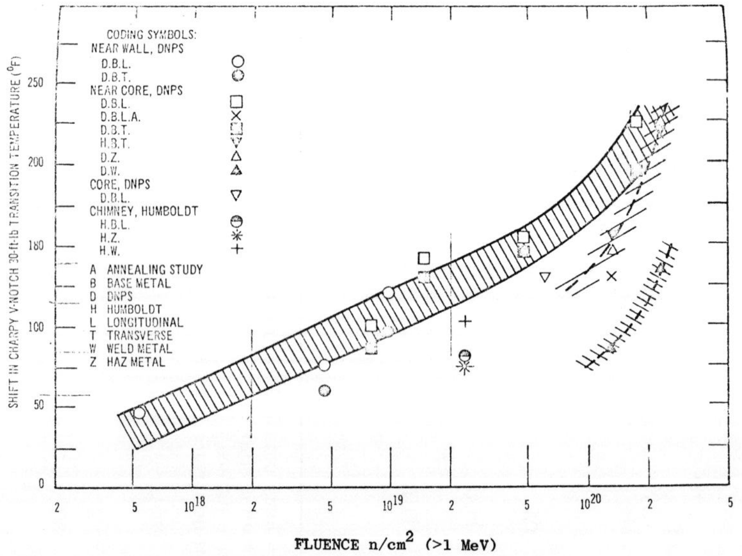

FIG. 93—*Increase in 30-ft·lb transition temperature versus neutaron fluence >1 MeV for DNPS-1 base, weld, and HAZ (Ref 19).*

A review of the DNPS surveillance Charpy curves reveals that the 30-ft·lb transition temperature increases and the upper energy absorption shelf is decreased as expected for the irradiation. No abnormal or potentially serious trends are observed. The upper shelf level of the HAZ specimens for the "two years near core" condition corresponding to a fluence of 2.3×10^{20} n/cm^2 >1 MeV is quite low, but, as will be seen in the subsequent section, this fluence is far in excess of the lifetime fluence level expected. Thus, the upper shelf decrease should be much less severe. Low energy tear would be of concern only if the extreme values for the HAZ were to be expected from vessel wall exposure.

The thermal control data (turning vane or steam drum) are interesting in that only the HAZ showed a decrease of the 30-ft·lb transition temperature, while the transverse base metal and the weld metal both showed an increase of about 25 F. Furthermore, the HAZ specimens revealed a lowering of the upper shelf energy, while the other two revealed higher upper shelf energy levels. The two-year exposure to temperature in the absence of radiation should be a reasonably severe test of the steel's ability to resist aging effects.

Tensile Properties

The tensile data from the DNPS surveillance program are summarized in Table 32 [*19*]. All testing was performed at 550 F on an Instron TT-C-L machine with a model FR-15,000-lb load cell. The crosshead speed was 0.050 in./min, and the strain rate 0.033 to about 0.025 in./min. For all materials, both the yield and tensile strengths increased as a result of irradiation, and the percent elongation plus percent reduction in area decreased although not to any large degree. In fact, this must be the most significant feature of the tensile data: although certain of the neutron fluences were far in excess of that expected for the 40-year service life, the base, weld, and HAZ all revealed very good retention of the important tensile ductility parameters. It should be noted, however, that the scatter of data within the material groupings is quite high.

Vessel Lifetime Projections

Neutron Dosimetry and Spectrum Considerations—The neutron flux and fluence values determined for the DNPS surveillance program were evaluated by both computer program and by foil dosimeter measurements. The calculated fluences were derived [*18*] from a computer program calculation of the Dresden reactor vessel using the APED 04-2 shielding code. This calculation was based upon the one-group removal method, which effectively gives fast neutron flux for energies of 0.1 MeV and greater. Interestingly, these values were generally lower than the foil measurements. The latter were based upon an assumed fission spectrum, as will be detailed in Chapter 5.

TABLE 32—*Average tensile for Dresden nuclear power reactor surveillance specimens* (*Ref* 19).

Material and Code	Irradiation Exposure				Integrated Neutron Flux Nvt > 1 MeV		Yield Strength, ksi	Tensile Strength, ksi	Elongation, (%)	Reduction in Area, %
	Group	Location	Years in Location	Equivalent Years (at wall)	Calculated	Measured				
Base o		control	...	0	...	...	58.2	83.0	24	70
	5	inside wall	4	6.4	2.9×10^{18}	9.8×10^{18}	68.9	93.9	22	69
Base oo		control	...	0	...	...	55.4	82.9	21	59
	5	inside wall	4	6.4	2.9×10^{18}	9.8×10^{18}	66.7	92.2	21	80
	12	in core	1	~ 75	...	6.1×10^{19}	74.8	95.3	17	45
Base ooo		control	...	0	...	...	53.6	83.0	25	69
	5	inside wall	4	6.4	2.9×10^{18}	9.8×10^{18}	66.0	93.1	23	70
Base oooo		control	...	0	...	...	60.2	87.2	24	63
	5	inside wall	4	6.4	2.9×10^{18}	9.8×10^{18}	69.5	94.1	20	75
oo	2	turning vane	2	0	$\sim 3 \times 10^{14}$	$<2 \times 10^{14}$	59.3	84.3	23	60
Standard oo	10	near core	1	~ 150	...	1.3×10^{20}	80.5	96.0	19	51
(Humboldt)	11	near core	2	~ 250	...	2.3×10^{20}	90.8	108.7	18	42
Weld DW		control	...	0	...	...	75.0	89.4	20	63
	2	turning vane	2	0	$\sim 3 \times 10^{14}$	$<2 \times 10^{14}$	74.4	88.3	19	66
	10	near core	1	~ 150	...	1.3×10^{20}	84.1	90.3	23	69
	11	near core	2	~ 250	...	2.3×10^{20}	99.1	104.0	17	61
HAZ DH		control	...	0	...	...	62.6	80.1	19	61
	2	turning vane	2	0	$\sim 3 \times 10^{14}$	$<2 \times 10^{14}$	63.2	85.1	18	76
	10	near core	1	~ 150	...	1.3×10^{20}	76.3	89.2	14	44
	11	near core	2	~ 250	...	2.3×10^{20}	89.9	104.3	15	73

An interesting point to note is that the cross-section and half-life values used to convert ^{54}Mn activity into flux were given in the first Dresden Ref *18* but were omitted in the second [*19*]. It is worthy of mention because the initial values were 53 mb and 300 days in contrast to the currently accepted cross-section value which ranged from 68 to almost 100 depending upon the specific laboratory, and a half-life of 314 days. It would appear that the old constants, used in Ref *18* were altered for use in the second publication as some of these initial fluence values have been reduced in the latter. Unfortunately, there is no record of values used in the second Ref *19*. Knowledge of the cross-section values used is most important, since without that knowledge it must be assumed that the same constants were used to determine all of the fluence values published in Ref *19* wherein the data of this report are taken.

The very large amount of surveillance data from DNPS-1 surveillance program should permit a clearcut estimate of the lifetime vessel condition. Unfortunately, this does not appear to be the case for Dresden. The reason is that a flux value for the vessel wall cannot be determined with confidence from the available data in the two references. Unless an accurate vessel wall flux is known, it cannot be extrapolated to service lifetime.

Reference *19* contains a table, reproduced as Table 33, that summarizes the exposure days of the surveillance capsules and yields a 40-year "calculated" flux value of 2×10^{19} n/cm^2 >1 MeV for the reactor. Using the fluences of Table 31 (measured values) and the exposure times of Table 33, it should be possible to arrive at flux values for the individual capsule exposures and extrapolate them to the vessel wall flux level through use of the constants provided. As will be shown later, the "measured" flux values derived from the data of Tables 31 and 33 do not all agree with the "calculated" flux and fluence values shown on Table 31 and a 40-year service fluence can be projected as much higher than shown in Table 33.

Consider the Group 3 capsule exposure first because it does agree with Table 33. From Table 31 note that base metal Group 3 capsules were irradiated near the wall for one year to a fluence of 0.5×10^{18} n/cm^2 >1 MeV. (Reference *18* gives this fluence as 5.8×10^{17} n/cm^2 >1 MeV.) The one-year base metal specimens were withdrawn in October 1961; thus, the specimens were exposed for time periods: 1, 2, and 3 (from Table 33). The 183 days exposure equals 1.58×10^7 s. Then:

$$\frac{5.0 \times 10^{17}}{1.58 \times 10^{7}} = 3.16 \times 10^{10}\ \text{n/cm}^2\text{/s} > 1\ \text{MeV}.$$

This is the value at the surveillance location. The flux at this location is 1.6 that at the vessel wall [*19*]. Accordingly, wall flux is:

$$\frac{3.16 \times 10^{10}}{1.6} = 1.98 \times 10^{10}\ \text{n/cm}^2\text{/s} > 1\ \text{MeV}.$$

TABLE 33—*Calculated fast neutron flux (nvt >1MeV) (Ref* 19).

Time Period	Dates	MWe-Days (approximate)	Equivalent Days[a]
1	April 1960	2 160	13
2	June 1960 to Nov. 1960	10 200	61
3	June 1961 to Oct. 1961	18 300	109
4	Dec. 1961 to Nov. 1962	54 400	323
5	March 1963 to April 1964	55 200	328
6	June 1964 to March 1965	40 600	242
7	May 1965 to Oct. 1966	...	500[b]
8	1967 to 1970	...	1200[b]
9	1971 to 1980	...	3000[b]
10	1981 to 1990	...	3000[b]
11	1991 to 2000	...	3000[b]

Location	Time Period	Equivalent Days[a]	n/cm²-s >1 MeV	Integrated nvt >1 MeV	Cumulative nvt >1 MeV
Near core 159 cm from core center	1 + 2	74	7.8×10^{11}	5.0×10^{18}	0.5×10^{19}
	3	109	1.0×10^{12}	9.6×10^{18}	1.5×10^{19}
	4	323	1.0×10^{12}	2.9×10^{19}	3.4×10^{19}
	5	328	1.0×10^{12}	2.9×10^{19}	6.3×10^{19}
	6	242	1.0×10^{12}	2.1×10^{19}	8.4×10^{19}
Between wall and shield (near-wall) 180.5 cm from core center	1 + 2	74	2.4×10^{10}	1.5×10^{17}	0.2×10^{18}
	3	109	3.2×10^{10}	3.0×10^{17}	0.5×10^{18}
	4	323	3.2×10^{10}	8.9×10^{17}	1.4×10^{18}
	5	328	3.2×10^{10}	9.1×10^{17}	2.3×10^{18}
	6	242	3.2×10^{10}	6.7×10^{17}	2.9×10^{18}
At wall ID 185 cm from core center	1 + 2	74	1.5×10^{10}	1.0×10^{17}	0.1×10^{18}
	3	109	2.0×10^{10}	2.1×10^{17}	0.3×10^{18}
	4	323	2.0×10^{10}	5.6×10^{17}	0.9×10^{18}
	5	328	2.0×10^{10}	5.6×10^{17}	1.4×10^{18}
	6	242	2.0×10^{10}	4.2×10^{17}	1.9×10^{18}
	7	500[b]	2.0×10^{10}	8.6×10^{17}	2.7×10^{18}
	8	1200[b]	2.0×10^{10}	2.1×10^{18}	4.8×10^{18}
	9	3000[b]	2.0×10^{10}	5.2×10^{18}	10×10^{18}
	10	3000[b]	2.0×10^{10}	5.2×10^{18}	15×10^{18}
	11	3000[b]	2.0×10^{10}	5.2×10^{18}	20×10^{18}

[a] Days of operation at 80 percent load factor; (0.8) (210) = 168 MWe with a 464 element core.
[b] Estimated.

Again, this value agrees with the "At Wall I.D. 185 cm from Core Center" fluxes of Ref *19*.

Consider the base metal Group 4 exposures now. These specimens were irradiated for two years to a fluence of 4.7×10^{18} n/cm^2 >1 MeV (Table 31). From Table 33, it can be taken that the two-year base metal specimens were withdrawn after time period 4 in November 1962. The total 506-day exposure period translates into 4.37×10^7 s. Then:

$$\frac{4.7 \times 10^{18}}{4.37 \times 10^7} = 1.075 \times 10^{11} \text{ n/cm}^2\text{/s} > 1 \text{ MeV}.$$

Using the same factor for flux at the vessel wall,

$$\frac{1.075 \times 10^{11}}{1.6} = 6.7 \times 10^{10} \text{ n/cm}^2\text{/s} > 1 \text{ MeV}.$$

Note that this value is significantly higher than that developed for the one year near-vessel exposure.

An examination of the notch ductility data adds validity to the greatly increased flux value for the two-year exposure. Note in Table 31 that the Group 3 fluence was 0.5×10^{18} but that the fluence for Group 4 was 4.7×10^{18}. This is not a threefold increase but a $(3 + 3)$ increase. Recall from Table 33 that the exposure time for Group 3 was 183 days, and for Group 4 was 506 days; this is close to a threefold increase in exposure time, and would account for an increase in fluence from 0.5 to 1.5×10^8. If now the vessel flux level were increased threefold, as is suggested by the one-year versus two-year flux values determined above, then the 1.5×10^{18} flux would be transformed to 4.5×10^{18} n/cm^2/s which essentially duplicates the measured value of 4.7×10^{18}.

The paradox of estimating Dresden flux value can be seen from further tabulating the flux values derived for other surveillance exposures:

Group 5	1,076 days	$1.05 \times 10^{11}/1.6$	$= 6.59 \times 10^{10}$ n/cm^2/s >1 Mev
Group 6	73 days	$1.11 \times 10^{12}/50$	$= 2.22 \times 10^{10}$ n/cm^2/s >1 MeV
Group 7	183 days	$9.5 \times 10^{11}/50$	$= 1.9 \times 10^{10}$ n/cm^2/s >1 MeV
Group 8	506 days	$1.10 \times 10^{12}/50$	$= 2.2 \times 10^{10}$ n/cm^2/s >1 MeV
Group 9	1,076 days	$1.88 \times 10^{12}/50$	$= 3.77 \times 10^{10}$ n/cm^2/s >1 MeV

The data of Groups 6 through 9 were from accelerated irradiation locations; the flux was taken [*19*] to be 50 times greater at these locations than at the vessel wall.

Projection—The vessel wall extrapolation flux values should be used for service limit projections because the vessel wall surveillance locations are physically closer to the reactor vessel and should not be far off from the 1.6 factor assigned to them. By the same argument, the accelerated location capsules are quite far from the core, and, in conclusion, the extrapolation factor of 50 was not determined rigorously and, in fact, would appear from the computed flux values to be only approximate.

If the vessel wall extrapolation flux is taken as being realistic, then we also must choose the Groups 4 and 5 flux values because they complement the fluence values determined by ^{54}Mn dosimetry. Accordingly, then, it is possible to determine an estimated service time from Table 33 to be 11,776 days which then will translate an average vessel wall flux of 6.6×10^{19} n/cm^2 >1 MeV.

It is now of high interest to determine the significance of this value upon the trend of increasing radiation embrittlement versus fluence for the DNPS-1 shown in Fig. 93. The fluence of 6.7×10^{19} n/cm^2 >1 MeV crosses the upper limit of the band at an increase in transition temperature of about 175 F. When added to the preirradiation transition temperature of +90 F for the transverse orientation, a maximum vessel transition temperature of 265 F is predicted for the 40-year service time of the DNPS-1. Using the NDT plus 60 F criterion, the minimum temperature for pressure tests at the 40-year point in service would be 325 F. This compares rather favorably with the 290 F value determined for the 2×10^{19} n/cm^2 >1 MeV fluence shown in Table 33.

Summary

It should be recalled that the entire Dresden vessel analysis had to be built up from transition temperature increase data representative of the steel from the vessel plate surface layers not the quarter thickness layers as is standard practice. Because no data exists concerning the characteristics of the inner layers of the Dresden vessel steel, no solid projections can be made for either the quarter or half thicknesses. An analysis for the Yankee reactor vessel (above) was possible only because such data were in the literature.

The trend of increasing transition temperatures for the DNPS base metal is more severe than either the weld metal or HAZ. None of the materials are what would be considered as highly sensitive to neutron radiation. However, it should be noted that the surface steel, because of improved microstructure, may show less embrittlement than would be expected from longitudinal $1/4T$ specimens. This low sensitivity is fortunate since it appears that the flux at the DNPS vessel wall and thus the lifetime fluence is relatively high. The initial steel transition temperatures being quite high, of course, work in opposition to the apparent low sensitivity of the DNPS vessel steel.

Despite the fact that the notch ductility of the DNPS vessel steel has shown a significant decrease, the tensile ductility, although also decreased, has not been drastically lowered. The increase of 25 F in the 30-ft·lb transition temperature of the DNPS weld and base metal due solely to temperature effects is somewhat surprising. However, the specimens at the beltline actually represent the point of concern, if any, as they undergo both thermal and nuclear embrittlement. If the two are cumulative, no matter; the specimens still tell the story at the worst point and little is gained from the thermal controls for projecting a specific reactor condition. As higher alloy steels and higher temperatures are used, however, the chance for embrittlement due to thermal aging grows as does the potential concern for this factor. For A302-B and A533 steels, thermal controls would seem to be a waste. Future Dresden results, nevertheless, will be of interest to test this analysis.

Elk River Reactor

Surveillance Program

The Elk River reactor (ERR) is a 58.2 MWt boiling water plant designed by Allis-Chalmers for the Rural Cooperative Power Association and is located in Elk River, Minn. The 3-in.-thick pressure vessel was fabricated of ASTM A302-B steel, internally clad with 0.109-in. Type 304 stainless steel. The discovery of cracks in this stainless steel cladding was the principal cause for the initiation of a very large-scale surveillance program. Despite the comprehensive nature of the program which is described in Ref *20*, however, no steel from the actual ERR pressure vessel was available for use in the irradiation studies.

Materials—The materials employed for the ERR surveillance program include three heats of A302-B steel, two heats of A105 Grade II steel and one heat of A212-B steel. An additional heat of A105-II steel in the form of a trepanned flange dropout also was included since it was the only material likely to be representative of the nozzle material. The chemistry and sources of ERR surveillance materials are listed in Table 34; the code identifications in the table will be retained for purposes of this analysis.

Data reported by Allis-Chalmers [*21*] and confirmed by the Southwest Research Institute (SwRI) [*22*] indicated that the A302-B alloy 3-1 was in the optimum heat treated condition and may not be fully representative of the material that had gone through vessel fabrication operations. To provide a wider spectrum of materials for the surveillance program, specimens were machined from alloy 3-2 which had been strained and stress relieved to simulate the shell rolling operation used in the ERR vessel and from alloy 3-3, a 3-in.-thick Lukens clad plate similar to the ERR pressure vessel material. This latter plate (Heat No. A5498) was obtained from the Argonne National Laboratory and had been used during initial ERR investigation. It was reported to have been manufactured at the same time as the ERR vessel and had a very poor notch toughness in the as-received condition [*20*].

TABLE 34—*ERR surveillance program materials* (*Ref* 20).

Code	Alloy	Source	Chemical Analysis, weight %						ASTM Grain Size
			C	Mn	P	S	Si	Mo	
1–1	A105–II	Allis–Chalmers[a]	not available						7
1–2	A105–II	SwRI[b]	0.26	0.73	0.008	0.023	0.22	...	4 to 5
1–3	A105–II	ERR (vessel flange[c]	not available						5
2–1	A212–B	Hanford[d]	0.29	0.80	0.015	0.023	0.26	...	8
3–1	A302–B	Allis–Chalmers[a]	not available						8 to 10
3–2	A302–B	SwRI[e]	0.22	1.32	0.014	0.021	0.24	0.46	8
3–3	A302–B	Argonne[f]	not available						8 to 10

[a] Initial surveillance program material.
[b] 1 by 6 by 24-in. plates in as-forged condition.
[c] Trepanned from flange by Allis–Chalmers.
[d] U. S. Steel Heat No. 73M 573, Plate No. 252696.
[e] 3-in. plate stock used in effect of defects program at SwRI [AEC Contract AT (11–1)–1228], strained and stress relieved.
[f] 3-in. Lukens clad plate, Heat No. A4116, used in Elk River investigation.

TABLE 35—*Drop-weight and Charpy V-notch properties of ERR surveillance program materials* (*Ref* 20).

Code	Alloy	Drop-Weight NDT	Charpy Correlation Temperature
1-1	A105-II	+37 F	+70 F (20 ft·lb)
1-2	A105-II	. . .	+42 F (20 ft·lb)
1-3	A105-II	. . .	+65 F (20 ft·lb)
2-1	A212-B	. . .	+10 F (20 ft·lb)
3-1	A302-B	−83 F	−63 F (30 ft·lb)
3-2	A302-B	. . .	−47 F (30 ft·lb)
3-3	A302-B	. . .	+77 F (30 ft·lb)

The notch toughness of the A105-II alloy 1-1 was slightly better than that of alloy 1-3 which had been removed from the vessel flange, and the microstructures were also different. Insufficient material of alloy 1-3 was available although it may possibly have been more representative of the nozzle forgings than alloy 1-1 and would have been considered desirable for inclusion in the surveillance program. Another heat of A105-II alloy 1-2, with a microstructure similar to and a notch toughness slightly inferior to that of alloy 1-3, was included in the surveillance program [*20*].

The A212-B steel was obtained from the reference steel plate mainained at the Battelle-Northwest Laboratory at Richland, Wash.

The available pre-service NDT and transition temperature properties of the ERR surveillance material are shown in Table 35.

Only the A302-B steel of the core region vessel plates will be of immediate concern in terms of radiation-induced increase of transition temperature, and it has been pointed out that no material in the surveillance program is from the vessel itself, or can be known to have the same characteristic radiation response of the vessel steel. In addition, the pre-service Charpy V-notch and tensile properties of the actual vessel steel are not known. In the absence of conclusive measurements, an initial worst-case transition temperature was assumed to be +100 F for the ERR vessel [*22*]. Analysis of the ERR vessel then must be based upon the development of an acceptable radiation response behavior trend from the characteristic ERR neutron and temperature environment with a superimposed initial steel transition temperature of +100 F.

Surveillance Capsules and Locations—Twenty of the initial 40 ERR surveillance capsules contained Charpy V-notch specimens of alloy 3-1, the A302-B supplied by Allis-Chalmers; tension specimens of alloy 3-1 were in twelve of the remaining capsules; and alloy 1-1 Charpy V-notch

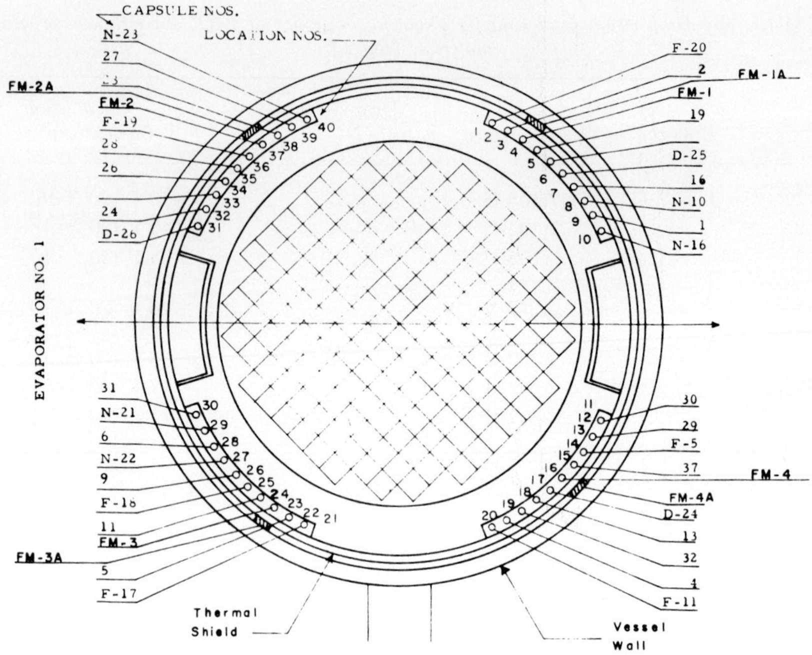

Fig. 94—*Surveillance capsule locations in Elk River reactor* (*Ref* 20).

and tension specimens were loaded into the other eight. As the initial capsules were withdrawn for testing, they were replaced with new compound capsules which contained both Charpy V-notch (or tension) and fatigue specimens. Charpy V-notch and fatigue specimens for the replacement capsules were machined from alloys 3-2 and 1-2. Additional fatigue specimens were machined from a dissimilar weld configuration of A105-II and Type 304 stainless steel simulating the 16-in. nozzle-to-cap weld made in the field and left in the as-welded condition. This dissimilar weld represents an out-of-flux material condition.

The 40 ERR surveillance capsule locations are shown schematically in Fig. 94. Each of these locations is an accelerated irradiation site as they are between the thermal shield and core. Four flux monitor capsules recently have been placed close to the vessel wall, however. These are identified in Fig. 94 as FM-1A, -2A, -3A, and -4A and are to be compared to results from accelerated locations FM-1, -2, -3, and -4 to verify the calculated flux difference factor of 2.3 [*20*] between the ERR accelerated surveillance location and the vessel wall.

The importance of the flux decrease from the accelerated location through the vessel wall flux monitor location to the alloy steel/stainless steel clad interface should not be minimized because it is this factor which will dictate the relative level of radiation embrittlement projected for the ERR vessel. In assessing the future measured neutron flux values from all surveillance locations, it will be important to determine accurately the relationship of the vessel wall flux monitors to the alloy steel/stainless steel clad interface as the flux will be reduced further over this distance. Two of the four pairs of flux monitor capsules (sets 1 and 3) are facing the same configuration of fuel elements so that if successive core loadings are similar, the vessel flux values should correspond. The other two sets of monitor capsules are facing altogether different core configurations and will yield different results. The values relating to measurements nearest the corners of the core (FM locations 1 and 3) will yield the highest and hence the limiting vessel fluxes.

Notch Ductility Properties

ERR surveillance capsules were withdrawn on two occasions. Capsules removed at the initial withdrawal had fluences of 0.22, 0.83, and 1.04 × 10^{19} n/cm² >1 MeV; fluences of 1.48 and 1.73 × 10^{19} n/cm² >1 MeV were recorded for the second withdrawal. Using the factor of 2.3 to relate surveillance capsule fluence with vessel wall fluence and 8.52 × 10^{18} n/

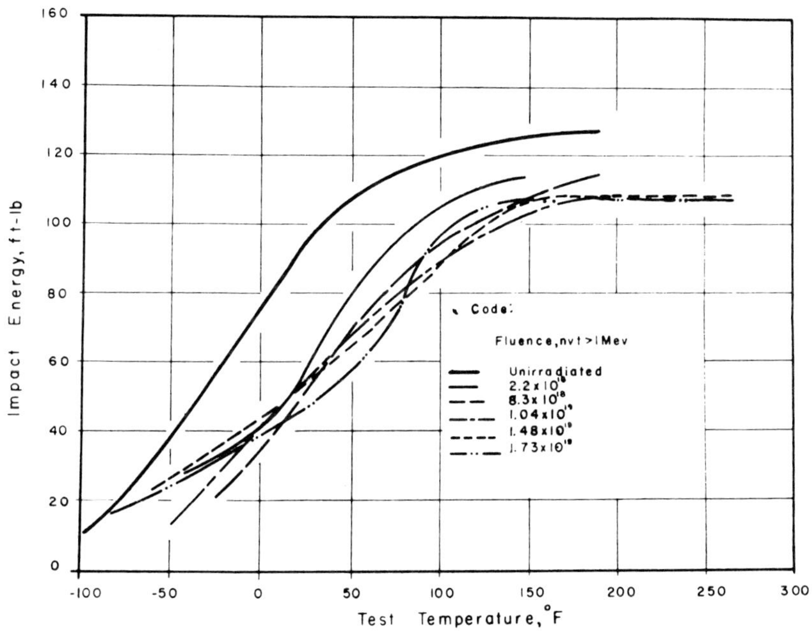

FIG. 95—*Charpy V-notch ductility characteristics of Elk River reactor surveillance steel alloy 3-1 before and after irradiation* (*Ref* 20).

cm^2 >1 MeV being equivalent to one full power year, the above-noted fluences correspond to 0.6, 2.25, 2.8, 4.0, and 4.7 full power vessel years of life. For reference, the reactor only had operated for about two years at the time of the second withdrawal.

The results of evaluating Charpy V-notch specimens of alloy 3-1 from capsules representing each of the five fluences are summarized in Fig. 95 [*20*]. The legend shows that a contradictory pattern of transition temperature increase with neutron fluence was observed. The exceptional similarity of all five curves would suggest that alloy 3-1 is closely approaching a condition of radiation embrittlement saturation for the temperature, the steel, and the nuclear environmental conditions of the irradiation locations or that the fluences are in error. For reference, the 550 F response of ASTM A302-B steel from the "reference" plate would amount to an increase of about 140 F for a 2.1×10^{19} n/cm^2 >1 MeV fluence. It would seem appropriate, therefore, to presume that alloy 3-1 displays a radiation response atypical of the usual A302-B reactor vessel plate. In fact, it is among the more insensitive commercial heats studied to date.

Charpy V-notch results from two fluence level exposures of alloy 3-2, the strained and stress relieved alloy, are shown in Fig. 96. The 30-ft·lb transition temperature increases for these exposures appear to be more in line with the response defined by the A302-B reference steel, but another anomaly exists in that both exposures, 0.83 and 1.48×10^{19} n/cm^2 >1 MeV, appear to have effected about the same ΔT (~130 F).

Alloy 3-3, the 3-in. clad plate from Lukens, displayed 20 and 40 F increases for exposures of 0.48 and 0.83×10^{19} n/cm^2 >1 MeV, respectively. For this one alloy only, the transition temperature increases were in order with the fluence received but were quite low relative to the behavior of the A302-B reference steel for similar fluences.

Transition temperature increases for A212-B steel, alloy 2-1, were measured along the 20-ft·lb energy level. The 50 F increase for 1.48×10^{19} n/cm^2 >1 MeV does not compare well with the 55 F increase determined for an 0.83×10^{19} n/cm^2 >1 MeV exposure. The unusually large number of these anomalous responses suggests either that a generalized error (such as in dosimetry or temperature during exposure) has crept in or that they represent actual, not anomalous, behavior. It is possible that specific combinations of capsules from locations closer or further from the fuel core with the attendant differences in neutron distribution are the cause. The location of the individual capsules with respect to the fuel core is not known to the author, but, what may be more important, neither are there available reasonably accurate estimates of the neutron energy distribution at these locations.

Embrittlement Projection—All of the transition temperature increases measured by SwRI for the ERR surveillance materials have been plotted

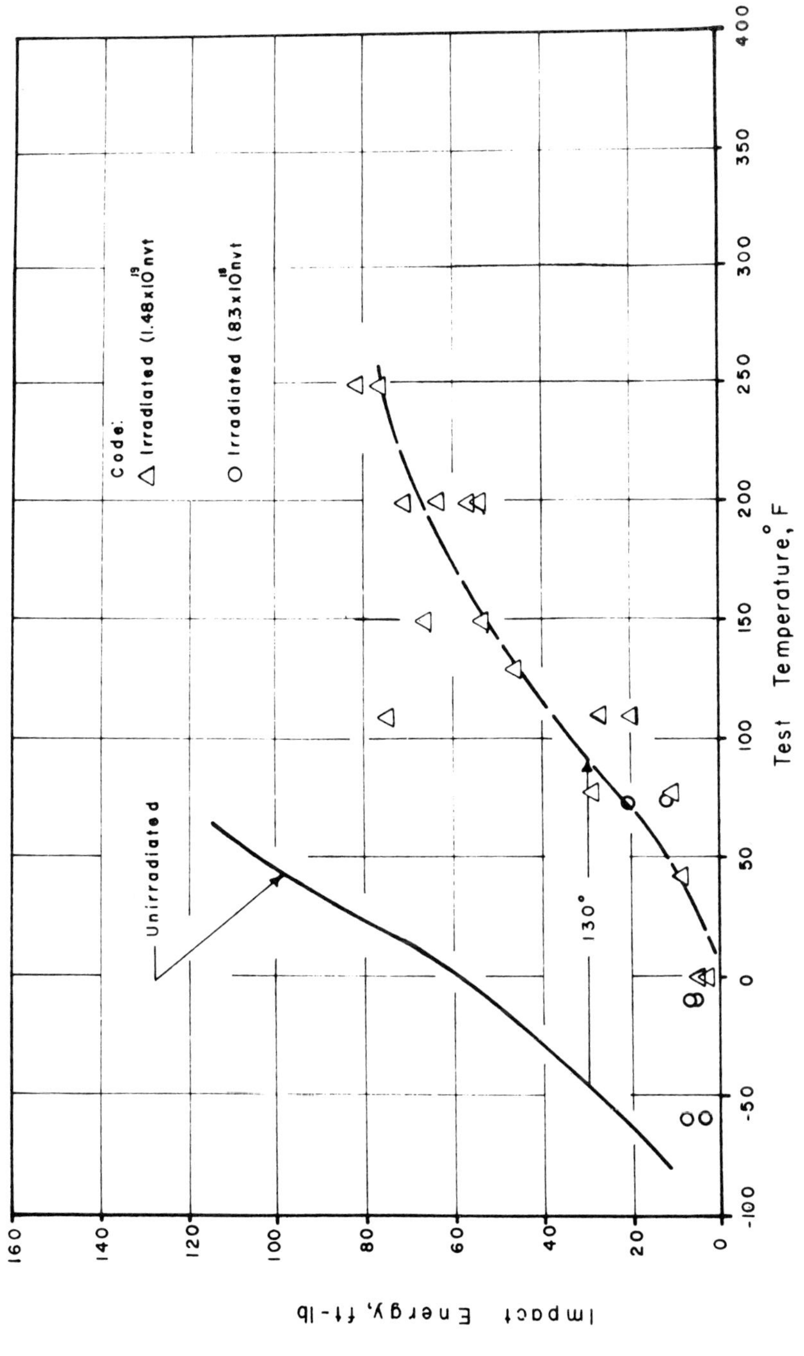

FIG. 96—*Charpy V-notch ductility characteristics of Elk River reactor surveillance steel alloy 3-2 before and after irradiation (Ref 20).*

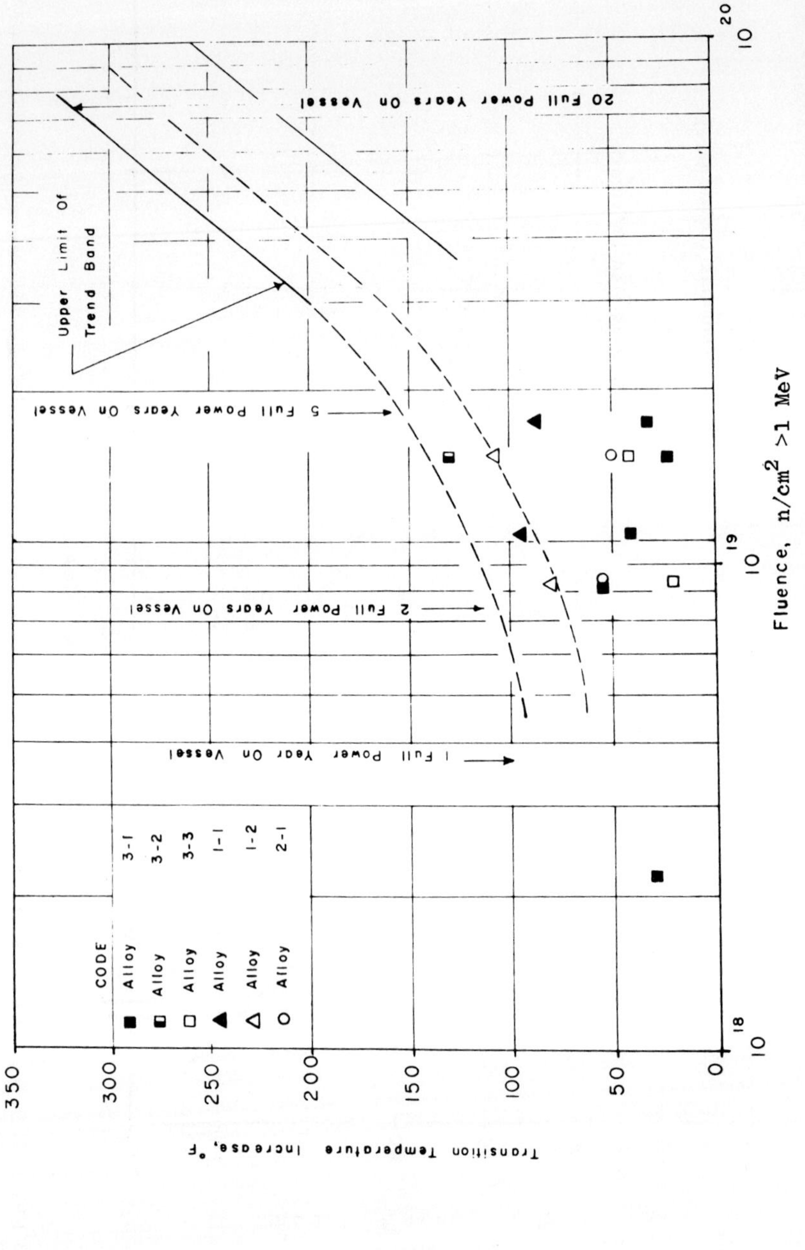

FIG. 97—*Increase in NDT versus fission spectrum neutron fluence >1 MeV for Elk River reactor surveillance steel (Ref 20).*

versus neutron fluence >1 MeV in Fig. 97 [*20*]; these are referenced to the NRL trend band for 550 F irradiation of A302-B reference steel. At the end of two full power years, the predicted shift in transition temperature of the vessel wall (at the core centerline) is 105 F or less. Similarly, after 5 full power years, a 150 F or less shift is expected, and after, 20 full power years, a shift in the transition temperature of 320 F or less is predicted. The transition temperature shifts obtained from the Charpy impact surveillance specimens are considerably less than the above predictions, possibly the response of unique materials in a unique environment [*20*].

Now, consider the data on a most conservative basis. If 320 F is taken as the maximum increase for the ERR vessel and to it is added the assumed initial 100 F transition temperature, there emerges a lifetime maximum transition temperature for the ERR vessel of +420 F prior to application of the NDT plus 60 F criterion. Such a value suggests the need for steps to prevent significant pressurization at below NDT especially during startup and shutdown. These problems would be expected to arise long before the 20 year reactor lifetime period is reached.

An additional problem in determining a reasonable radiation response for the ERR vessel steel is that of defining the irradiation temperature. The ERR vessel apparently operated between 519 and 536 F, but the surveillance capsules were subjected to some gamma heating and gave melting point monitor temperature indications of between 516 and 604 F. Certainly 550 F is a close approximation; but if the actual surveillance temperature were less, as indicated, then a certain amount of annealing would have been assumed by the 550 F prediction and the embrittlement would be understated. However, with the actual irradiation surveillance capsules it is possible that specimens were exposed above 550 F. In this case, some of the apparent anomalies might be explained by the higher exposure temperatures. The influence of temperature above 550 F is most significant as evidenced by a recent experiment conducted in the BRPR. In this experiment, a number of steels irradiated at the reactor ambient of about 585 F exhibited markedly low ΔT values for the fluences involved. For example, the reference A302-B shift was 165 F for an exposure of 7.1×10^{19} n/cm^2 >1 MeV, essentially the same shift as that projected for the ERR exposure of only 2.1×10^{19} at 550 F.

Tensile Properties

Tension tests were performed on six miniature and one standard tension specimen of alloy 3-1 and on six miniature tension specimens of alloy 1-1. The average tensile properties of the irradiated specimens are compared to those of the unirradiated specimens in Table 36. The initial neutron exposure of 2.2×10^{18} n/cm^2 >1 MeV increased the tensile strength and decreased the tensile ductility. Although the tensile ductility has con-

TABLE 36—*Average room-temperature tensile properties of irradiated specimens compared to the unirradiated condition (Ref 20).*

Alloy	Exposure, nvt >1 MeV	Gage Length, in.	Ultimate Tensile Strength, ksi	0.2% Yield Strength, ksi	Yield Strength / Ultimate Tensile Strength	Elongation, %
3-1	none	1.0	94.3[a]	63.5[a]	0.674	35.0[a]
3-1	2.2×10^{18}	1.0	108.6	87.6	0.807	24.5
3-1	1.04×10^{19}	1.0	107.6	86.6	0.806	24.5
3-1	1.73×10^{19}	1.0	98.5	78.4	0.796	19.9
1-1	none	1.0	67.4[a]	43.2[a]	0.641	37.6[a]
1-1	1.04×10^{19}	1.0	87.6	57.4	0.654	22.2
1-1	1.73×10^{19}	1.0	76.8	55.7	0.725	21.8

[a] Allis-Chalmers data from ACNP-64582.

tinued to decrease with additional exposure (to 1.73×10^{19} n/cm^2 >1 MeV), the tensile strength properties went through a maximum value at a point between 2.2×10^{18} and the unirradiated values. A satisfactory explanation for these anomalies has not yet been developed. Again, lack of knowledge of temperatures during exposure hampers the analysis. Based on present data, however, the radiation-induced changes in tensile properties do not seem to be even potentially troublesome.

Fatigue Properties

For the ERR surveillance program, fatigue strength was defined [*20*] as the stress to cause failure at 10,000 cycles in an unnotched specimen that is subject to constant load fatigue. The surveillance results to date showed that the fatigue strength was increased from as little as 15 percent to as much as 30 percent for both alloys 1-2 and 3-2. The dissimilar welds were affected the least but still indicated an increase in fatigue strength. Based on a constant cyclic load condition, irradiation has enhanced the fatigue strength of these materials [*20*].

This conclusion is interestingly in conflict with experimental in-reactor fatigue results developed by NRL on a section of 3-in.-thick A302-B steel supplied by SwRI as representative of ERR stock. The experiment, which has been described [*23*], was conducted for a relatively low fluence of about 8×10^{18} n/cm^2 >1 MeV at 550 F. Sheet type specimens fatigued in reverse bending mode were studied for both the irradiated and unirradiated conditions. The results, shown in Fig. 98, indicate no evidence

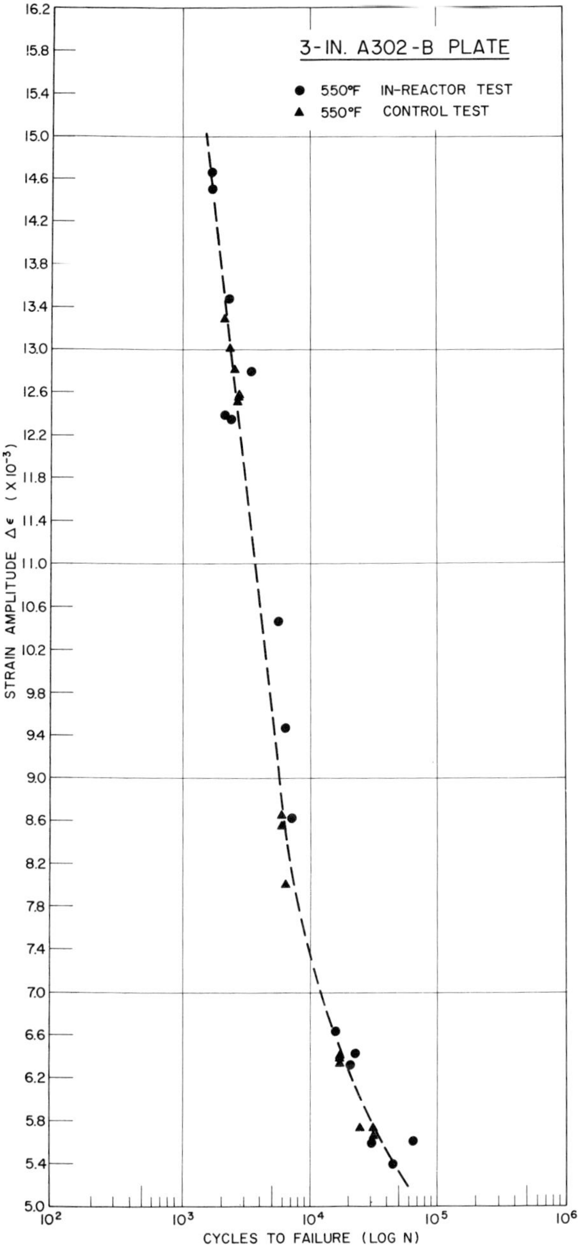

FIG. 98—*Comparison of fatigue data for the A302-B plate developed by in-reactor tests at 550 F with the results of out-of-reactor control tests. No evidence of an irradiation effect is apparent (Ref 23).*

of an irradiation effect for the conditions and material of the experimental irradiation. It is possible that these specimens experiencing a slightly higher temperature, 550 F, underwent some annealing of the radiation damage or that cycling during exposure permitted correction of damage more readily. By comparison, the ERR surveillance fatigue specimens were irradiated at a lower temperature, about 536 F, under harder neutron spectrum conditions to a higher fluence and underwent no cycling during exposure. Nevertheless, a conclusive explanation for the differing fatigue results is not apparent at this time.

Summary

By far the most important conclusion that can be drawn from the ERR surveillance program is that quantity of surveillance materials cannot make up for a lack of truly representative material. Even the inclusion of an extensive tension and fatigue specimen program cannot rescue the overall program results because these efforts also are directed to unrepresentative materials.

One of the brighter aspects of the program, however, is the effort to measure the neutron flux at both the vessel wall and the surveillance locations. This will provide an important value which will supersede the calculated vessel flux value, if it can be interpreted in terms of the neutron spectral distribution at the vessel wall. Irradiation results from test reactor irradiations eventually will be used to predict the ERR vessel lifetime, but such projections cannot be considered valid without knowledge of the vessel wall neutron spectrum.

If the projection of a 20-year embrittlement of 320 F with an initial steel transition temperature of 100 F are shown to be accurate, then the ERR would be in potential difficulties in the later stages of its lifetime for startup and shutdown pressurization situations. Thus, current dosimetry and surveillance efforts will be of prime importance in providing facts for a firm analysis of ERR vessel conditions in the future.

Humboldt Bay Reactor

The Humboldt Bay reactor (HBR), a unit of the Pacific Gas and Electric Company, Humboldt Bay Power Plant, is a 240 MWt natural circulation boiling water plant having a 5-in.-thick pressure vessel of A302-B steel. The data summarized for the irradiated surveillance materials are taken from Ref *24*. The chemical composition and heat treatment of the vessel, shown in Table 37, were taken from Ref *25* which details fabrication history of the vessel and surveillance steel.

Surveillance Program

The steel used for the HBR surveillance program was from the plate-forming stubs of three plates from the HBR reactor vessel; these stubs

TABLE 37—*Chemical composition and heat treatment of Humboldt Bay reactor pressure vessel surveillance steel* (*Ref* 25).

Lukens Heat Number	Vessel Location	Lukens Analysis, weight %					
		C	Mn	Si	P	S	Mo
20192-2	shell at core	0.20	1.27	0.21	0.016	0.036	0.47
20192-3	shell at core	0.20	1.32	0.18	0.020	0.023	0.46
A0638-2	shell above core	0.20	1.25	0.22	0.020	0.029	0.45

Heat Treatment

Austenitized at 1600 F for 4 h, brine quenched; tempered at 1225 F for 4 h, furnace cooled. Longitudinal vessel welds adjacent to core were exposed to 1150 F stress relief treatments for an accumulated 20 h. The surveillance steel was given a simulated postweld heat treatment; subsequently cut specimen blocks were stress relieved at 1150 F for an additional 10 h followed by furnace cooling.

were integral with the shell-course plates through the heat treating operations and cold forming into half shell courses. Postweld stress reliefs were applied to the three plates to simulate actual vessel conditioning. Two stubs from core shell-course plates were used for preparation of Charpy V-notch and tension specimens. The stub from the shellcourse above the core was used to prepare the weldment for the HAZ and weld specimens.

Capsules for specimens of the HBR were identical to those for the Big Rock Point reactor. The surveillance program provided for capsules containing Charpy V-notch and tension specimens of base metal, HAZ, and weld metal at accelerated locations (designated Chimney), at vessel wall locations, and at out-of-flux locations for monitoring thermal effects alone. Sufficient capsules were provided to permit three separate withdrawals for all materials and both specimen types from the Chimney locations, four withdrawals from the vessel wall locations, and three withdrawals from thermal control locations. Spare capsules in thermal control locations are also available for three additional withdrawals.

After two years operation of the HBR, surveillance capsules were removed from a Chimney location to yield Charpy V-notch and tension specimens of base metal, HAZ, and weld metal. The irradiation temperature of the specimens, as confirmed by inspection of low-melting point temperature monitors, was about 545 F, the reactor operating temperature. The measured neutron fluence for the exposure was reported to be 2.85×10^{19} n/cm^2 >1MeV based upon a 60-mb cross section for the ^{54}Fe (n, p) ^{54}Mn reaction [*24*]. As all fission spectrum fluences in this report are normalized to 68 mb for the iron reaction cross section, the HBR value

TABLE 38—*Summary of Charpy V-notch ductility results for the Humboldt Bay reactor pressure vessel surveillance program (irradiation in Chimney location at 545 F to 2.52 × 10*^{19} *n/cm*^{2} *>1 MeV* (*Ref* 24).

Material	Charpy V-notch 30-ft·lb Transition Temperature, deg F			Absorption at Full Shear, ft·lb	
	Unirradiated	Irradiated	ΔT	Unirradiated	Irradiated
Base	0	80	80	95	80
HAZ	−20	55	75	60	45
Weld	−25	80	105	88	68

has been recalculated and, accordingly, will be considered to be 2.52 × 10^{19} n/cm^{2} >1 MeV.

Notch Ductility Properties

In line with the Dresden reactor surveillance program practice, but contrary to that in the Big Rock Point reactor, the HBR Charpy specimens were taken from the plate surface layers. The longitudinally oriented base metal specimens were notched perpendicular to the plate surface which is common practice, but weld metal specimen notches were cut parallel to the plate surface. HAZ specimen notch roots carefully were aligned to lie on the weld edge as revealed by etching; the notches were cut parallel to the plate surface.

The Charpy V-notch results for base metal, HAZ, and weld metal specimens from the initial surveillance withdrawal of the HBR are summarized in Table 38 [*24*]. The base metal and HAZ specimens showed 30-ft·lb transition temperature increases of 80 and 75 F, respectively, with the final transition temperature residing at or below 80 F. The weld metal specimen transition temperature also resided at 80 F after irradiation, but this was the result of a 105 F increase. Thus, the weld metal exhibited the highest rate of transition temperature increase.

Further consideration of the HBR Charpy V-notch results reveals that the decrease in energy absorption at full shear fracture is approximately equivalent for all three materials. Although the postirradiation shelf levels for the base metal and weld metal specimens are between 70 and 80 ft·lb, the HAZ specimens exhibited a postirradiation shelf level of about 45 ft·lb. This low shelf level does not appear to result from scatter since the scatterband of about ±10 F was quite consistent over the 300 F testing range in contrast to the wide degree of scatter exhibited by the HAZ tension specimens. It should be recognized that the upper shelf energy absorption for the HAZ Charpy specimens is low enough to be of concern, especially since there is good evidence to support the fact that they represent well the property they are designed to monitor. Thus, the HAZ properties of the HBR emerge as the potential weakest link (limiting material) in the vessel weldment.

Tensile Properties

The 1 1/2-in. gage length 0.250-in.-diameter specimens were all tested at 550 F. The results are summarized in Table 39 [*24*]. Both the yield and tensile strengths increased as a result of the irradiation. The percent reduction in area for all materials showed a decrease in total elongation (in. 1.5 in.); however, only the irradiated HAZ specimens clearly showed a decrease. The fluence of 2.5×10^{19} n/cm^2 >1 MeV is sufficient to effect some reduction in tensile ductility of A302-B steels, but such does not appear the case for the HBR steel.

The difficulty in placing the gage center directly on the fusion line for best HAZ definition plus the probability for off-center breaking appears

TABLE 39—*Summary of Humboldt Bay reactor surveillance tensile data from 550 F tests (Ref* 24).

Material Type	Yield Strength, ksi	Ultimate Strength, ksi	Uniform Elongation in 1.5 in., %	Total Elongation in 1.5 in., %	Reduction Reduction in Area, %
Control:					
Base..........	59.07	80.46	9.3	17.5	59.5
Base..........	59.28	81.07	9.5	18.1	62.5
Base..........	60.90	81.99	9.5	18.0	64.5
Weld..........	55.00	75.37	7.5	14.5	63.5
Weld..........	60.30	78.42	7.3	15.2	67.0
Weld..........	57.24	75.37	7.3	15.3	67.0
HAZ..........	58.87	76.59	8.8	16.7	68.0
HAZ..........	65.69	81.99	9.0	17.9	67.0
HAZ..........	61.11	78.02	8.8	17.3	58.5
Irradiated:					
Base..........	73.33	93.70	10.8	18.0	46.0
Base..........	73.33	93.91	10.3	17.7	53.0
Weld..........	73.33	91.66	7.6	14.3	55.0
Weld..........	72.31	90.65	9.9	17.3	56.5
HAZ..........	68.24	90.14	7.9	14.0	43.0
HAZ..........	74.35	92.28	8.6	15.7	64.5

to be evident from the scatter in those values. Both of the other two materials, however, produced internally consistent results.

Summary

The longitudinally oriented base metal tension specimens of HBR steel irradiated in the Humboldt Bay reactor displayed strength increases reasonably commensurate with the incident neutron fluence and surprisingly good retention of tensile ductility characteristics. Transversely oriented base metal tension specimens of HBR steel also were irradiated in accelerated locations of the Dresden-I Reactor. The results of these irradiations [*19*] compare favorably with those from the HBR, particularly with respect to the trend of tension ductility retention. In the referenced irradiations, a fluence of 1.3×10^{20} n/cm^2 >1 MeV produced a lowering of the reduction in area for transversely oriented specimens from 63 to 49 percent. As an overall assessment, the tensile properties of the Humboldt Bay reactor steel do not appear to suggest the potential for any future problems based upon present knowledge.

The 30-ft·lb transition temperature increases for the HBR surveillance Charpy V-notch specimens are compared above in Fig. 93 to other Charpy V-notch increases for both Humboldt Bay and Dresden reactor vessel steels. The band defines, principally, the increases from Dresden (DNPS) steel irradiated in the Dresden-I reactor. (For reference, the Big Rock Point increases were very similar in magnitude to those of the HBR but were from an irradiation to a slightly higher fluence.) An assessment of the HBR Charpy V-notch impact results based upon Fig. 93 would have to suggest that relatively small transition temperature increases should accrue to the reactor during a 40-year life. This, of course, is based upon the premise [*24*] that the fluences reported here are far in excess of the expected lifetime fluence.

The Charpy V-notch specimens involved represented steel from the top layer of the vessel plate but not the quarter-thickness layer which is used for plate qualification. As indicated for Dresden, data from surface layer steel cannot be compared fairly to trends developed using quarter-thickness steel. Nevertheless, this steel layer will receive the maximum flux, be irradiated the most, and may thus better represent the limit condition for vessel embrittlement.

The upper shelf energy values for the HAZ Charpy V-notch specimens are of concern for the future. If, as it is suggested [*24*], the fluence of 2.5×10^{19} n/cm^2 >1MeV is in excess of that to be expected for the vessel, then the reduction noted also should be in excess of that to be expected for the vessel. Accordingly, it could be projected that the upper shelf will not be reduced to a sufficiently low enough level to be susceptible to low energy tear failure. As has been recommended [*24*] however, a set of vessel wall surveillance specimens should be withdrawn at the five-year interval to check on the upper shelf energy degradation and to obtain a valid and much needed neutron flux and fluence measurement.

Concern over the 40-year fluence value for the HBR is justified principally because no measurement has yet been made at the vessel wall. It is true that a value for the accelerated (Chimney) location has been determined [24], but it is noted that this reference states that a flux extrapolation to the vessel wall can be off by a large factor, thus necessitating the capsule removal to permit a direct measurement. A comparison of the three General Electric boiling water reactors (Big Rock Point, Dresden, and HBR) reveals no obvious pattern, probably because each has a vessel and a core configuration different from the others. Nevertheless, judging from the core layouts with respect to the vessel walls of both the BRPR and HBR, and using the projected fluence values for the BRPR, it would appear that a 40-year operating period for the HBR will yield a low enough fluence so that problems arising from deleterious mechanical property changes cannot be reasonably foreseen; however, this conclusion will be verified or denied fortunately by the results of future surveillance tests.

References

[1] Whitman, G. D., Robinson, G. C., Jr., and Savolainen, A. W., "Technology of Steel Pressure Vessels for Water-Cooled Nuclear Reactors," ORNL-NSIC-21, Oak Ridge National Laboratory, Dec. 1967.

[2] "Recommended Practice for Surveillance Tests on Structural Materials in Nuclear Reactors," ASTM E 185-70, *1970 Book of ASTM Standards*, Part 30.

[3] Brandt, F. A. and Alexander, A. J., "Effect of Neutron Irradiation at 550 F on Charpy Impact Properties of ASTM A302 Grade B Steel," *Radiation Effects on Metals and Neutron Dosimetry, ASTM STP 341*, American Society for Testing and Materials, 1963, pp. 212–232.

[4] Brandt, F. A., "Reactor Pressure Vessel Materials Surveillance Program at the Consumers Power Company Big Rock Point Nuclear Plant," General Electric Company, GECR-4442, Dec. 1963.

[5] Serpan, C. Z., Jr., Steele, L. E., and Hawthorne, J. R., "Radiation Damage Surveillance of Power Reactor Pressure Vessels," *Transactions*, American Society of Mechanical Engineers, Series D, Vol. 89, No. 1, March 1967, pp. 221–232.

[6] Hawthorne, J. R. et al, "Irradiation Effects on Reactor Structural Materials, Quarterly Progress Report, 1 Aug.–31 Oct. 1967," NRL Memorandum Report 1833, Naval Research Laboratory, 15 Nov. 1967.

[7] Potapovs, U. and Hawthorne, J. R., "The Effect of Residual Elements on 550 F Irradiation Response of Selected Pressure Vessel Steels and Weldments," NRL Report 6803, Naval Research Laboratory, 22 Nov. 1968.

[8] Serpan, C. Z., Jr., et al, "Yankee Reactor Pressure Vessel Surveillance: Evaluation of Specimens Exposed During the Second Core," NRL Report 6179, Naval Research Laboratory, 24 Nov. 1964.

[9] Landerman, E. A., "Surveillance Tests on Structural Materials in Nuclear Reactors," *Radiation Effects on Metals and Neutron Dosimetry, ASTM STP 341*, American Society for Testing and Materials, pp. 233-253.

[10] Serpan, C. Z., Jr., Steele, L. E., and Hawthorne, J. R., "Radiation Damage Surveillance of Power Reactor Pressure Vessels," NRL Report 6349, Naval Research Laboratory; 31 Jan. 1966, also *Transactions*, American Society of Mechanical Engineers, Vol. 89, Series D, No. 1, March 1967, pp. 221–232.

[11] Serpan, C. Z., Jr., and Hawthorne, J. R., "Yankee Reactor Pressure Vessel Surveillance: Notch Ductility Performance of Vessel Steel and Maximum Service Fluence Determined from Exposure During Cores II, III, and IV," NRL Report 6616, Naval Research Laboratory, 29 Sept. 1967.

[12] Steele, L. E. et al, "Irradiation Effects on Reactor Structural Materials, Quarterly Progress Report, 1 May–31 July 1965," NRL Memorandum Report 1638, Naval Research Laboratory, 15 Aug. 1965.

[13] Carpenter, G. F., Knopf, N. R., and Byron, E. S., "Anomalous Embrittling Effects Observed During Irradiation Studies on Pressure Vessel Steels," *Nuclear Science and Engineering*, Vol. 19, 1964, pp. 18–38.

[14] Steele, L. E., Hawthorne, J. R., and Watson, H. E., "Irradiation Effects on Reactor Structural Materials, Quarterly Progress Report, 1 May-31 July 1963," NRL Memorandum Report 1448, Naval Research Laboratory, 15 Aug. 1963.

[15] Steele, L. E. et al, "Irradiation Effects on Reactor Structural Materials, Quarterly Progress Report, 1 May–31 July 1968," NRL Memorandum Report 1908, Naval Research Laboratory, 15 Aug. 1968.

[16] Serpan, C. Z., Jr., and Watson, H. E., "Notch Ductility, Tensile, and Neutron Spectrum Analyses of PM-2A Reactor Pressure Vessel," NRL Report 6739, Naval Research Laboratory, 19 June 1968; also *Irradiation Effects in Structural Alloys for Thermal and Fast Reactors, ASTM STP 457*, American Society for Testing and Materials, 1969, pp. 135–155.

[17] "Yankee Plant–Fast Neutron Exposure, Transition Temperature Increase, and Operating Limitations for the Reactor Vessel," Yankee Atomic Electric Company Report RFS-M-5, June 1966.

[18] Brandt, F. A. and Alexander, A. J., "Dresden Nuclear Power Station Reactor Vessel Steel Surveillance Program," APED-3988, Atomic Power Equipment Department, General Electric Co., July 1962.

[19] Hersh, M. S., Brandt, F. A., and Beaudreau, B. C., "Dresden Nuclear Power Station Reactor Vessel Steel Surveillance Program," GECR-5165, Atomic Power Equipment Department, General Electric Co. May 1966.

[20] Ireland, D. R. and Norris, E. B., "Influence of Neutron Irradiation on the Properties of Steels and Welds Typical of the ERR Pressure Vessel After Two Power Years of Operation," SwRI 1228P9-15, Southwest Research Institute, March 1968.

[21] MacKay, D. S., "Unirradiated Vessel Material Nil-Ductility Transition Test," ACNP-64582, Allis-Chalmers Manufacturing Co., 1964.

[22] Norris, E. B., Southwest Research Institute, private communication, 20 Sept. 1968.

[23] Hawthorne, J. R. and Steele, L. E., "In-Reactor Fatigue Tests of A302-B Steel Representing the Elk River Reactor Pressure Vessel," NRL Memorandum Report 1596, Naval Research Laboratory, 15 Jan. 1965.

[24] Brandt, F. A., "Humboldt Bay Power Plant Unit No. 3, Reactor Vessel Steel Surveillance Program," GECR-5492. Atomic Power Equipment Department, General Electric Co., May 1967.

[25] Brandt, F. A., "Reactor Pressure Vessel Material Surveillance Program at the Pacific Gas and Electric Company Humboldt Bay Power Plant, Unit No. 3," GECR-4443, Atomic Power Equipment Department, General Electric Co., San Jose, Calif. Dec. 1963.

CHAPTER 5—NEUTRON DOSIMETRY AND SPECTRUM

Introduction	209
Neutron Dosimetry	209
Activity	210
Cross Sections	212
Neutron Flux	214
Neutron Spectrum	217
Spectral Computer Codes	218
Description of Spectral Plots	220
Application of Spectral Data to Reactors	223
Big Rock Point Reactor Spectrum Analysis	225
Yankee-Rowe Spectrum Analysis	230
SM-1A—LITR Reactor Spectrum Analyses	233
Conclusions	234
References	234

CHAPTER 5—Neutron Dosimetry and Spectrum

Introduction

Both neutron dosimetry and neutron spectrum have been referred to extensively throughout this report. They are of equal importance with mechanical property changes since the changes would be essentially meaningless without the point of reference provided by dosimetry and spectrum. Just as creep rupture studies have an accurate time base and carefully documented environments, so must radiation effects studies have accurate neutron dosimetry for a base and accurate spectral analysis for environmental definition. Furthermore, a very important requisite for both fluence measurements and spectral definition is a damage model which relates experimentally measured values to real physical conditions in the reactor. Therefore, damage model considerations will be carried along throughout the discussion of the two principal subjects, dosimetry and spectrum.

In this chapter a brief, simplified review of the essential elements of neutron dosimetry is provided. Additionally, an introduction to the concept of neutron spectrum calculations is provided, and a simplified means for using and interpreting spectra is presented. Different proposed damage models also are noted and discussed. Finally, the step-by-step calculational procedure used to obtain the service life fluence values for two reactors is presented in both a tabular form and in the accompanying text.

Neutron Dosimetry

In its simplest form neutron dosimetry is a means for measuring the number of neutrons which are present and moving at a given location in a nuclear reactor.

Neutron flux, the basic unit of measure, is referred to more properly as flux density according to the latest recommendations of the International Committee on Radiation Units (ICRU). Nevertheless, the term "flux" has been and will continue to be used throughout this chapter as a convenience. Flux ($n/cm^2 \cdot s^{-1}$) is the product of the neutron density (n/cm^3) times the neutron velocity (cm/s). This flux, ϕ, appears in the activation Eq 1 which relates the induced activity of a neutron flux detector (or foil) resulting from reactor irradiation at saturation with its cross section for activation and the actual flux of the reactor.

$$A = N_0 \bar{\sigma} \phi \qquad (1)$$

where A is the saturated activity of the flux detector in disintegrations per second, N_0 is the number of atoms in the detector, $\bar{\sigma}$ is the activation cross section averaged over the neutron spectrum under consideration, and ϕ is the total number of $n/cm^2 \cdot s^{-1}$ in the same spectrum. A saturated activity simply means that the detector has been irradiated for a sufficient time period that new activated species are being produced as rapidly as previously produced species are decaying; that is, the activity of the detector material at this time under irradiation is constant.

A more general equation for activation is:

$$A = N_0 \bar{\sigma} \phi (1 - e^{\lambda t}) \qquad (2)$$

where λ is the half-life of the produced isotope, and t is the irradiation time. A detailed discussion of neutron flux calculations and interrelations is given by the ASTM Standard Method for Measuring Neutron Flux by Radioactivation Techniques (E 261–70 [*1*].[1]

The flux (ϕ) is the average value for the irradiation of interest and, thus, may be integrated over the total time period (t) to yield $\phi \cdot t$, the fluence (until recently referred to as dose or exposure).

Activity

As a means of better understanding the features of neutron dosimetry, consider the elements of Eq 1. First is the activity of the detector, A, which is measured as radioactive disintegrations per milligram of irradiated materials per second. Specific materials such as iron or nickel and cobalt commonly are used for this purpose in irradiation experiments as "neutron flux monitors." In reality, they only indicate activity from which "flux" can be calculated. The nuclei of certain isotopes of these elements accept neutrons, thereby becoming transmuted to higher isotopes; as these isotopes subsequently decay, specific gamma rays are associated with the decay. Fortunately, these gamma radiations are characteristic for each isotope and thus readily can be identified using multichannel analyzers. The radiation or photons so emitted during decay have very narrow energy limits and result in photo peaks at these energy levels in the radiation spectrum measured by the analyzer equipment. The number of individual photons counted from interaction with a crystal detector correspond to disintegrations of transmuted atoms with their attendant gamma radiation. By totaling these distinctive disintegrations, it is possible to calculate a flux detector activity.

One of the principal reasons that certain materials such as iron or nickel

[1] The italic numbers in brackets refer to the list of references appended to this chapter.

and cobalt are used for "flux monitors" is that each of their characteristic radiation-spectra are composed of only one or very few well-defined and well-separated gamma ray energy peaks thus forming a "fingerprint" so that identification is positive and counting can be very precise. A further reason is that their half-lives are long enough ($\geq$ two months) to be useful for irradiations covering long time periods. A long half-life is important to preclude activity saturation early in the irradiation, thereby decreasing the detector's sensitivity to operating changes late in the irradiation. Also, a long half-life will assure the activity to remain at a high, statistically satisfactory level for a long enough time after experiment discharge to permit accurate counting. Finally, the physical form of these materials is quite compatible with surveillance capsule designs and operating temperatures.

The three materials mentioned previously—iron, nickel, and cobalt—make a good combination for radiation damage and surveillance studies. Iron, using the reaction $^{54}Fe(n, p)^{54}Mn$, is of high importance as a "fast flux" detector as it can be used as pure metal wire or can be analyzed selectively from the alloy steel of an irradiated test specimen (or even a drill chip of a decommissioned reactor vessel). Nickel, using the $^{58}Ni(n,p)^{58}Co$ reaction, is useful for providing a check against the accuracy of the iron measurement—some insurance against a counting error. An advantage of nickel is that its cross section is better known than is iron's. A serious drawback to the use of nickel, however, is that its activation product (^{58}Co) has a 5400 (very high) barn cross section for thermal neutrons. This exceptionally large cross section dictates that the ^{58}Co isotope produced by the detector reaction will itself be transmuted rapidly up to ^{59}Co and then to ^{60}Co which would thus lead the researcher to conclude that he had a far lower flux (based upon the remaining ^{58}Co) than was the actual case. A correction can be made if the thermal flux is known, as it will then be possible to account for the transmuted ^{58}Co.

Cobalt is used for thermal flux measurements through the reaction $^{59}Co(n, \gamma)^{60}Co$. It is activated readily but requires significant corrections for self shielding in heavier thicknesses (greater than several mils diameter). Therefore, it commonly is alloyed with aluminum as Al–0.1Co (commonly referred to as Co–Al) wherein no shielding corrections are necessary.

To obtain a thermal flux with cobalt wire, a bare Co–Al foil is exposed as nearly as possible in the same physical location as a cadmium-covered Co–Al foil. The bare foil is activated by neutrons of all energies; the cadmium cover, however, absorbs the thermal neutrons (and all neutrons of energies up to $\sim$0.4 eV) but permits all higher energy neutrons to activate the Co–Al foil within the cadmium cover. Subtracting the "cadmium-covered" results from the "bare" results yields the thermal flux. For short-term irradiation, 0.020-in.-diameter cadmium is sufficient, but for longer irradiations and higher thermal fluxes, greater thicknesses are necessary to assure the presence of enough cadmium to preclude "burn-out" of cadmium absorption capability and consequent thermal neutron leakage. The presence of a cadmium cover will create a thermal flux depression within a radius of about one inch. Of necessity, bare cobalt detectors therefore, should be located outside this sphere if most accurate measurements are to be made.

Thermal flux measurements can also be made using the "two foil" method [*2*, *3*, *4*], whereby cobalt and silver detectors are exposed to the flux simultaneously without any cadmium cover, hence no cadmium-induced thermal flux depression. The detectors commonly are used as aluminum alloys having 0.1 weight percent cobalt or silver. The silver reaction, $^{109}Ag(n, \gamma)^{110m}Ag$ is well suited to pressure vessel surveillance irradiations because the product half-life is in excess of 250 days.

Cross Sections

Returning to the discussion of factors in Eq 2, the average cross section is the factor which determines how many or how few atoms of a detector material will become transmuted at a given time. The unit of measure, the barn, is 10^{-24} cm^2, an area which can be thought of as the opaqueness of a material or that relative ability of a material to interact with a neutron passing through the square centimeter of interest. An excellent review of cross-section theory, measurements, and behavior is given by Hughes [*5*].

Large cross sections, multiples of barns, reflect the fact that many neutrons and atoms interact and that many atoms will become transmuted per second in a reactor flux. The cross section can be given as an average value for activation by all the incident neutrons in a given reactor flux. However, such an average value requires that the actual distribution of neutrons by energy levels in the reactor flux be known or assumed to conform to some known distribution. The cross section also can be given as a series of values for the neutrons in several groups, each bounded by specific energy levels. For use of these group-averaged cross sections, it is also necessary to know the distribution of neutrons by energy levels since the neutrons in each energy group are multiplied by the group cross sections to obtain one overall spectrum-averaged cross section.

A great deal of interest and concentrated effort currently surrounds data and research in cross sections because at once, they are of exceptionally high importance and of significant difficulty to measure. It is not, however, as difficult to measure *total* cross sections, it is only difficult to separate the portions caused by absorption, scattering, and the like. For the purposes of this report, we are interested in just one of these portions—the absorption cross section, since it is this reaction which results in transmutation which ultimately can be measured and converted into neutron flux and fluence values. The necessity for accuracy in cross sections should

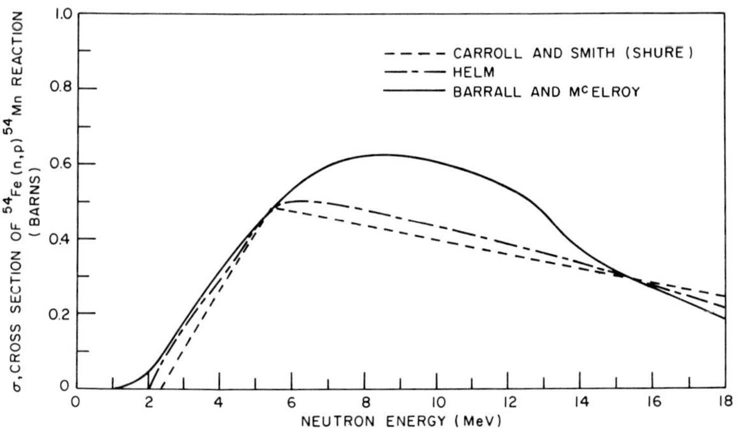

FIG. 99—*Cross-section measurements versus energy (MeV) of the* ^{54}*Fe* (n, p)^{54}*Mn reaction with the interpretation of the points by three invesigators* (*Refs* 6–8).

be apparent if one considers that a 25 percent increase in the cross-section value (as a result of new evaluations, for example) will result in a 25 percent decrease in the neutron flux and fluence for that particular reactor exposure.

Neutron flux derived from iron detectors is used for the neutron dosimetry values throughout this report. A short discussion of the activation cross-section measurements and values for the ^{54}Fe(n, p)^{54}Mn reaction is therefore in order. Accelerator studies have been carried out on iron to determine its activation as a function of energy. The data points from these accelerator studies, as well as the curves through the points by Carroll and Smith [*6*], Helm [*7*], and Barrall and McElroy [*8*], are shown in Fig. 99. The references noted on Fig. 99 are contained in Ref *8* of this chapter. Although the curves disagree significantly in the range between 6 and 14 MeV, there are few neutrons in a reactor spectrum in this range, so this disagreement will have little effect on the cross section. The really significant differences between the three curves are at the lower MeV portion of the curves, where the preponderance of activations occur. The effect of the curves in terms of activation cross sections for a fission spectrum is shown in Fig. 100. These response curves are obtained by multiplying the neutron population in a fission spectrum by the three cross-section measurements. As can be seen, there are wide differences in the portion of a fission spectrum considered by these three curves between 2 and 8 MeV, and particularly between 2 and 5 MeV. These differences give rise to the fission-spectrum averaged cross section for iron of 68 mb by Shure [*9*], from the work of Carroll and Smith, 82.6 mb by Helm, and 97.2 mb by Barrall and McElroy. The value of 82.6 mb determined by Helm has been

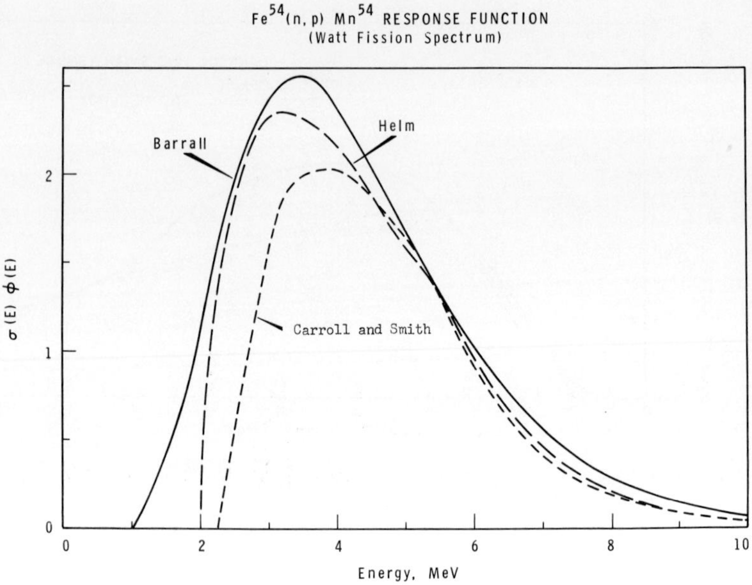

FIG. 100—*Response function of the* ^{54}*Fe* (n, p) ^{54}*Mn reaction to the Watt fission spectrum as influenced by the interpretation of three investigators.*

chosen as the fission-spectrum-averaged cross section for iron to be used for this report.

Neutron Flux

Cross sections must now be discussed together with the flux (Eq 1) since the two are highly interdependent. As has been indicated above, the flux is the actual number of neutrons present and moving at any location in a reactor radiation field. The flux (flux density) also may be thought of as the number of neutrons incident per unit time on a small sphere centered at that point divided by the cross-sectional area of that sphere. It is important to recognize that neutrons of all energy levels from thermal (between 0 and ~0.4 eV) up through fast (0.1 to about 18 MeV) are present in greater or lesser degree in every location of every reactor. The distribution of neutrons by energy levels is referred to as the neutron energy spectrum and will be discussed in greater detail in the following section. Because it is so very difficult to determine the neutron energy spectrum in a reactor, common practice has been to assume that the neutrons were distributed throughout the range of energies according to a fission neutron spectrum. One form of the fission spectrum, the Watt form, is shown in Fig. 101 as a histogram wherein the average energy level of the neutrons between energy limits is represented by the horizontal bars.

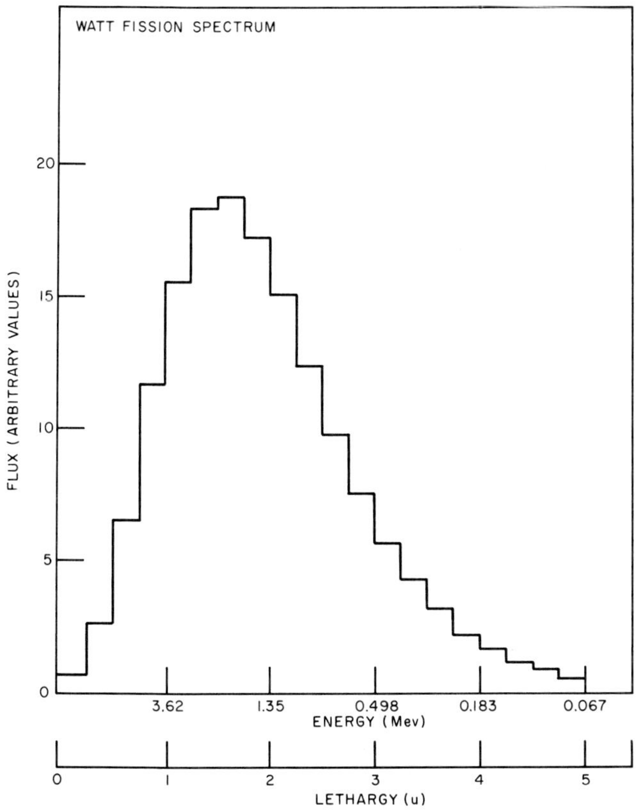

FIG. 101—*Graphical representation of the Watt fission spectrum plotted by arbitrary flux values in terms of* ϕ (u) *versus 0.25 lethargy* (u) *units* (*Ref* 10).

Several other forms of fission spectra also have been developed, particularly those of Cranberg [*10*] and Grundl [*11*], but the corresponding differences are not critical to this report. The fission spectrum in Fig. 101 is a plot of the flux (ϕ) in arbitrary units, as a function of their energy.

A short listing of neutron activation reactions, fission averaged cross sections, half-lives, and approximate effective threshold energies is given in Table 40 for some commonly used detectors in pressure vessel steel irradiation damage studies. The term threshold implies that no activation occurs below this energy level but that 100 percent activation occurs above this energy. Because it is so very difficult to determine the exact location of this threshold, it has been found convenient to use effective thresholds. Here, a point can be chosen relatively easily below which little activation will occur and above which most will occur. Although the amount of activation is always variable from the lowest energy of activation and up,

TABLE 40—*Critical factors for preferred threshold flux detectors.*

Detector Types	Detector Material	Nuclear Reaction	Fission-Averaged σ, barns	Approximate Threshold Energy, eV	Product Half-Life
Thermal	cobalt	$^{59}Co(n, \gamma)^{60}Co$	36.3	0.025	5.27 years
	cobalt (cadmium shield)	$^{59}Co(n, \gamma)^{60}Co$	36.3	0.414	5.27 years
Fission threshold	thorium	$^{232}Th(n, f)$	...[a]	1.73×10^6	...[a]
	neptunium	$^{237}Np(n, f)$	...[a]	0.75×10^6	...[a]
	uranium	$^{238}U(n, f)$	...[a]	1.45×10^6	...[a]
Fast threshold	iron	$^{54}Fe(n, p)^{54}Mn$	0.0826	$\sim 2.0 \times 10^6$	314 days
	titanium	$^{46}Ti(n, p)^{46}Sc$	0.0095	$\approx 3.0 \times 10^6$	84 days
	nickel	$^{58}Ni(n, p)^{58}Co$	0.0950	$\sim 3.0 \times 10^6$	71 days
	copper	$^{63}Cu(n, \alpha)^{60}Co$	0.00042	$>2.0 \times 10^6$	5.27 years

[a] Depends upon which fission product is selected. More than one is suitable for use.

as can be seen readily for the threshold detector iron in Fig. 99, in this case it is assumed that the activation takes place as a step function with activation occurring uniformly above that level. When neutron fluxes are stated for n/cm^2·s^{-1} "$>$1 MeV" it is here assumed that the activation occurs uniformly in a foil by all neutrons of all energies whose minimum energy level is 1 MeV. The activation cross section then simply is taken to be averaged over the assumed fission spectrum from 1 MeV and up.

It is at this point that significant problems occur in dosimetry measurements. Very simply, the problem is that only a small percentage of those neutrons having energies between 1 and 3 MeV are effective in activating iron. However, a very high proportion of the neutrons of energies $>$1 MeV exist between 1 and 3 MeV so that the iron detector reaction does not measure effectively the real number of neutrons present. In order to overcome this problem, it has been convenient in the past to assume that the number of neutrons between 1 and 3 MeV (and naturally of all other energies as well) in any reactor spectrum will be given by the fission spectrum distribution. It should now follow that if the actual numbers of neutrons between 1 and 3 MeV differ markedly from the prediction of the fission spectrum then the flux of neutrons calculated from an activated iron detector also will be in marked error. Although the neutron energy distributions in and near the cores of light water moderated test reactors are surprisingly similar to the fission spectrum, especially for neutrons $>$1 MeV, the neutron distributions near power reactor pressure vessels are decidedly not. Thus, if accurate measurements are to be made of the number of neutrons which impinge upon pressure vessel steels, particularly in reactor walls, then dosimetry must have available better documentation of the neutron energy distributions—neutron spectra.

Neutron Spectrum

A neutron spectrum, for the purposes of this report, is taken to mean the entire population of neutrons in the flux of a reactor location. For utility and ease of presentation, spectra will be tabulated as the population of neutrons whose energies reside between specified limits yielding about 21 groups, the entire range being from 0 to 10 MeV. Figure 101, the Watt form of the fission spectrum, has a number of features to be retained in the spectrum figures throughout this report. Notice that the abscissa has two scales, lethargy and MeV, and that the energy decreases from left to right. For orientation, this means that thermal neutrons (0 to ~0.4 eV) are far to the right, off the figure entirely. The lethargy scale is included since it is used by reactor physicists for spectral calculations in preference to an energy-based scale.

Fission spectra have been measured experimentally using elaborate experimental setups and, frequently, with enriched ^{235}uranium sources. It is pointed out that in no other place will such a spectrum exist, and hence other distributions must be employed.

The more common and rapid techniques in use today for calculations of neutron spectra use diffusion and transport theory as well as the Monte Carlo method to arrive at theoretical distributions of neutrons in given reactor physical arrangements. A relatively new technique called Spectral Analysis by Neutron Detectors (SAND–II) [*12*] shows great promise, however, for the most accurate definition of neutron spectra. This accuracy is attained by computerized curve-fitting using the activation results of between 10 and 30 different detector materials whose responses are overlapped sufficiently to be sensitive to neutrons of all energies; the detectors are irradiated simultaneously in the reactor location of interest. At present, the great value of SAND–II code and analysis system is for verification of the accuracy of the computer calculated neutron spectra.

Spectral Computer Codes

For any code it is first necessary to draw up a schematic layout representing the reactor location of interest. Frequently, this layout incorporates one quadrant of a reactor taken along lines of symmetry. Zones are set up that frequently correspond to fuel element locations and the nuclear densities, cross sections for scattering, etc., of all materials within that zone are tabulated and fed to the computer. The mathematical internals of the code then use this information to arrive at a solution which takes the form of populations of neutrons residing in groups of specified energy boundaries.

The choice of a type of spectral analysis computer code is dependent upon several factors. Among these is the overall size of the analysis problem. For relatively small problems such as in-core work, diffusion codes such as P1MG [*13*] are quite useful since they provide for rapid solutions and may not require significantly large correction factors. For larger problems, such as presented by the pressure vessel wall of a large power reactor, transport theory codes such as Program S [*14*] may be better utilized since they are mathematically better adapted for deep penetration problems. The P3MG Code [*15*] may have good utility if the problem is not too large since it contains transport theory modifications in a basically diffusion code; it thus combines relatively rapid solution with acceptable adaptation to deep penetration. Correction factors are still necessary for P3MG however.

As mentioned above, the speed at which a computer solution is reached is a factor to be considered. If total cost is critical, diffusion codes may be preferred since they generally require less computer time. Transport codes require longer computer times, and Monte Carlo methods may take even longer. It may be fair to say that for pressure vessel wall analyses, transport theory spectra appear to yield the most accurate results, although

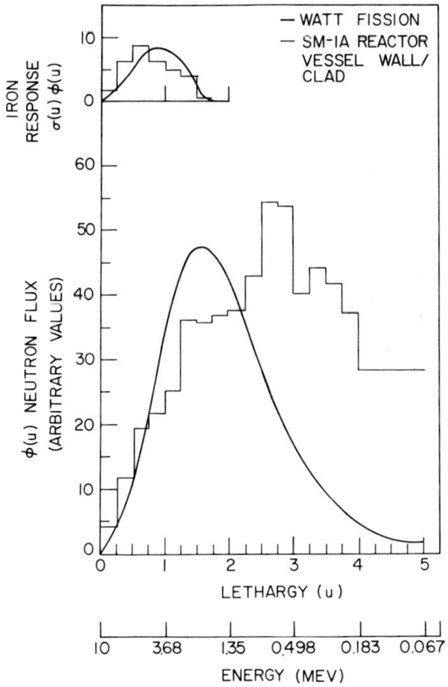

FIG. 102—*Neutron energy specturm of the Watt fission spectrum compared to a transport theory calculation of the SM-1A reactor vessel stainless steel clad interface.*

every reactor physics code now available was written for in-core analysis, not deep penetration analyses such as pressure vessel.

An example of the differences which can be obtained by using different computer code methods (but also, unfortunately, different libraries of cross-section values used as code input data) is illustrated by Figs. 102 and 103, that were adapted from Refs *16* and *17*. Here the interface of the alloy steel/stainless steel clad of the SM–1A reactor was calculated by the transport theory code, Program S in Fig. 102, and by the diffusion theory, P1MG code, in Fig. 103. The scheme used to plot these spectra will be explained below, but at this point note that both are related to the same magnitude of the Watt fission spectrum. It should be quite clear that the transport code, Fig. 102, has indicated the depletion of a great number of neutrons between about 0.1 and 4 MeV relative to the population shown for the P1MG spectrum. The points to be observed are that the number of neutrons >1 MeV in the diffusion spectrum, Fig. 103, is significantly larger than in the transport spectrum, but that the average neutron energy is significantly higher in the transport spectrum. For a given measured amount of radiation damage caused by the neutrons of this reactor, the

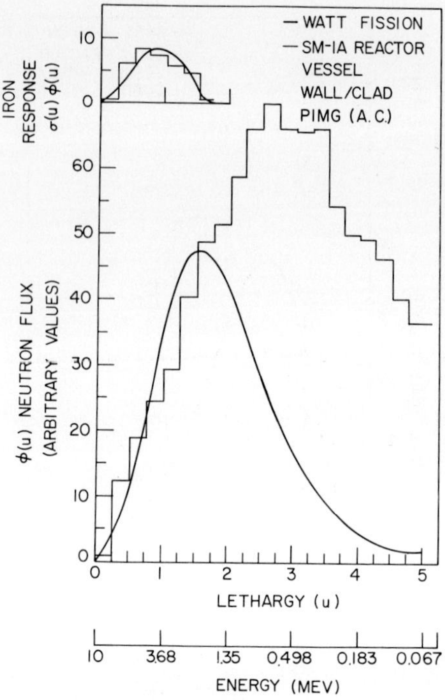

FIG. 103—*Neutron energy spectrum of the Watt fission spectrum compared to a diffusion code calculation of the SM-1A reactor vessel stainless steel clad interface.*

corresponding fluence quoted by reference to the P1MG spectrum would be significantly greater than the fluence quoted from the transport spectrum. Any lifetime fluence predictions for the reactor should be based upon the same computer code spectrum as used for development of the embrittlement trend. If radiation damage/fluence effects from other reactors, as a test reactor, were employed to establish the power reactor lifetime and the test reactor effects were based upon calculated spectrum fluences from a different code, the marked differences in relative neutron populations cited above could have a large effect upon the estimate.

Description of Spectral Plots

The spectra shown in Figs. 102 and 103 are typical of those to be presented and discussed throughout this chapter. The main portion of the figure is a composite of the Watt fission spectrum as the smooth curve, and the reactor spectrum of interest is presented as a histogram. The upper curves are the response of a neutron flux detector, iron in all the figures, to the flux in the fission spectrum compared to the response in the spectrum of the figure. The two curves in the lower portion relate directly to the

fluence quoted for an iron detector irradiated in the particular reactor if based upon either the fission spectrum or calculated spectrum criteria. For example, in Fig. 102, if n/cm^2 >1 MeV were cited, it should be apparent that the area under the fission spectrum is larger than the area under the calculated spectrum (1 MeV corresponds to a lethargy of about 2.3). Accordingly, if the fluence quoted for an irradiation at the SM–1A reactor vessel wall/clad (Fig. 102) were in terms of the fission spectrum, it would be overstating the number of neutrons actually present. If n/cm^2 >0.5 MeV were quoted, both the fission spectrum and calculated spectrum would be coincidentally very similar. But if n/cm^2 >0.1 MeV were quoted, the fission spectrum would greatly understate the actual population of neutrons. Since neutrons of energies well below 0.1 MeV contribute in some degree to radiation damage, use of the fission spectrum for fluence calculations clearly presents a false indication of neutron populations.

The smaller, upper curves in each figure are the key to the understanding and interpretation of the spectral plots. The two curves reveal the response of an iron detector to the neutrons in a fission spectrum and the spectrum of interest in the figure. Response is defined as the neutron flux population in each of the individual energy groups multiplied by the activation cross section for iron (the *group*-averaged cross section), for neutrons of those individual groups. In these upper curves then, the area under both curves are equal since both areas represent the entire activation of the iron detector. In the figures of this chapter, the group-averaged cross sections for iron are those constructed by Helm [*7*]. They are tabulated in Table 41 as a part of the listing of fluxes per energy and lethargy group resulting from a transport theory, Program S [*14*] calculation for three locations in the Big Rock Point reactor (BRPR). It should be recognized that starting with lethargy 2.00, which corresponds to energies less than 1.74 MeV, the cross section for iron activation is so small that it is essentially nil, meaning that none of the neutrons below this energy level can contribute to activation in an iron detector. This is precisely what the response curves reveal.

Not only do these response curves show the activation, they also describe how the activation is distributed. Refer back to the spectral plot of the BRPR, PVW/SSClad, Fig. 71. Note that the first three groups between lethargy 0 and 0.75 show neutron populations higher than the fission spectrum; consequently, the activation is less than that for a fission spectrum. Thus, it appears that neutrons of energies between 10 and 4.72 MeV in the BRPR PVW/SSClad spectrum account for a very large percentage of the activation of iron detectors compared to the activation within a fission spectrum. It is also reasonable to propose that these high energy neutrons will create a high proportion of damage. The importance of obtaining calculated neutron distributions in reactors such as the one shown for the BRPR should be quite apparent.

Because the response curves represent all the activation of an iron de-

TABLE 41—*Calculated neutron flux (arbitrary units) per energy group for the Big Rock Point reactor and Watt fission spectrum.*

Energy Group Structure			Calculated Relative Neutron Flux[a] $(n/cm^2 \cdot s)^{-1}$			Group Averaged Cross Sections	Relative Flux
Group No.	Lethargy (u)	Lower Energy Limit (E_L), eV	Accelerated Surveillance	Vessel Wall Surveillance	PVW/SS Clad Surveillance	^{54}Fe,[b] barns	Watt Fission Spectrum
1	0.25	7.78×10^{6}	1.234×10^{-7}	1.698×10^{-8}	1.266×10^{-8}	0.450	1.85×10^{-5}
2	0.50	6.07×10^{6}	3.928×10^{-7}	5.458×10^{-8}	4.104×10^{-8}	0.495	6.76×10^{-5}
3	0.75	4.72×10^{6}	6.776×10^{-7}	8.145×10^{-8}	6.167×10^{-8}	0.450	1.593×10^{-4}
4	1.00	3.68×10^{6}	8.019×10^{-7}	8.042×10^{-8}	6.203×10^{-8}	0.295	2.783×10^{-4}
5	1.25	2.87×10^{6}	9.376×10^{-7}	8.356×10^{-8}	6.493×10^{-8}	0.195	3.857×10^{-4}
6	1.50	2.23×10^{6}	1.481×10^{-6}	1.208×10^{-7}	9.158×10^{-8}	0.110	4.570×10^{-4}
7	1.75	1.74×10^{6}	1.125×10^{-6}	9.978×10^{-8}	8.228×10^{-8}	0.015	4.715×10^{-4}
8	2.00	1.35×10^{6}	1.088×10^{-6}	1.018×10^{-7}	8.745×10^{-8}		4.485×10^{-4}
9	2.25	1.05×10^{6}	8.694×10^{-7}	9.057×10^{-8}	8.598×10^{-8}		3.885×10^{-4}
10	2.50	8.21×10^{5}	7.907×10^{-7}	9.229×10^{-8}	9.558×10^{-8}		3.213×10^{-4}
11	2.75	6.39×10^{5}	8.899×10^{-7}	1.014×10^{-7}	1.038×10^{-7}		2.537×10^{-4}
12	3.00	4.98×10^{5}	7.470×10^{-7}	9.383×10^{-8}	1.082×10^{-7}		1.952×10^{-4}
13	3.25	3.88×10^{5}	5.141×10^{-7}	5.941×10^{-8}	6.032×10^{-8}		1.462×10^{-4}
14	3.50	3.02×10^{5}	6.277×10^{-7}	6.979×10^{-8}	6.403×10^{-8}		1.080×10^{-4}
15	3.75	2.35×10^{5}	5.553×10^{-7}	6.131×10^{-8}	5.657×10^{-8}		7.53×10^{-5}
16	4.00	1.83×10^{5}	4.922×10^{-7}	5.166×10^{-8}	4.222×10^{-8}		5.37×10^{-5}
17	6.75	1.17×10^{4}	3.452×10^{-6}	3.656×10^{-7}	3.225×10^{-7}		9.88×10^{-5}[c]
18	9.00	1.23×10^{3}	2.279×10^{-6}	2.335×10^{-7}	1.897×10^{-7}		
19	11.50	1.01×10^{2}	2.647×10^{-6}	2.522×10^{-7}	1.824×10^{-7}		
20	16.50	6.83×10^{-1}	5.916×10^{-6}	4.963×10^{-7}	3.798×10^{-7}		
21	20.72	1.0×10^{-4}	1.535×10^{-4}	7.180×10^{-6}	3.922×10^{-6}		
		$\phi(E_L > 1$ MeV)	7.655×10^{-6}	7.484×10^{-7}	6.078×10^{-7}		
		$\phi(E_L > 0.5$ MeV)	9.924×10^{-6}	1.017×10^{-6}	8.963×10^{-7}		

[a] The low values for the fluxes are only peculiarities of the computer program output. They are not to be construed as real flux values.
[b] Reference *4*.
[c] Lethargy for this value is 5.00.

tector in the two spectra under consideration in the figure, it follows that the area under both response curves must be equal. This, then, provides a simple basis for establishing a scheme to normalize the fission spectrum and spectrum of interest. To obtain the areas using Table 41, for example, sum the products of group flux times their group average cross sections, Next, adjust one total to the other to obtain a difference factor, and then multiply the group fluxes by this factor. In this way, both spectra can be plotted on the same set of axes, and they will be in proportion to each other exactly as described by their responses for either the fission spectrum or reactor spectrum. If spectra were plotted so that the areas under the fission spectrum and spectrum of interest were equal, it would be possible to obtain a false impression of the relative importance of neutrons of different energies. By equating the response curve, the true differences between quoting fluences from the fission spectrum and the calculated spectrum are apparent.

Application of Spectral Data to Reactors

Three schemes for use of neutron spectral data in the interpretation of radiation effects are now considered. These include one by Shure [*9*], one by Rossin [*18*], and one described in different applications by Serpan and Steele [*16*] and Serpan and Hawthorne [*19*] at NRL.

The method of Shure is based upon the establishment of a set of energy-dependent damage cross sections for neutrons of energy levels between 67 keV and 10 MeV, not just for the first seven energy level groups, as was described above for activation cross sections in connection with Table 41. The calculated spectrum fluxes are multiplied by the damage cross sections, a sum is taken and used as the "fluence" for plotting versus radiation damage parameters. Such a procedure is far more realistic than simply accounting for neutrons supposed to be in a fission spectrum. The specific values of the damage cross sections are not absolutely correct but are the best theoretical values presently available. Therefore, the resultant fluences values obtained will be in error only to the same extent as are the cross sections. The fluence values produced by this method are referred to as "*Dt*" meaning time-integrated damage and are usually in the range of 10^{21} to 10^{23} for the type of exposure usually associated with about 10^{19} n/cm^2 >1 MeV. In order to perform this type of calculation, it is necessary to normalize the neutron spectrum above 67 keV by the ratio of calculated to detector-measured activity.

The method of Rossin also employs a set of damage cross sections which are the basis for some of Shure's cross sections. These are also for multiplication by group flux values determined from spectral calculations. Rossin, however, introduces the concept of a Radiation Damage Unit (RDU) which is closely tied to the ^{54}Mn production rate. He defines RDU thus: 1000 neutrons having a perfect fission-spectrum energy distribution will deliver 1000 RDU [*18*]. The same comments concerning accuracy of damage cross sections and normalization of the spectrum apply here. And again, the technique used by Rossin is superior to that of simply assuming a fission spectrum distribution for fluence determinations. Rapid comparisons of irradiation effects data based upon the RDU concept (and that of *Dt*) and data based upon the far more common n/cm^2 concept, however, are not easily made.

Another, more recent, damage model analysis technique has been presented by Sheely [*20*]. This work also is concerned with relating the displacement density in a reactor spectrum to radiation damage. Sheely, however, has considered all of the following possible interactions: anisotropic elastic high-energy neutron scattering, inelastic high-energy neutron scattering, thermal-neutron capture-gamma recoil-induced displacements, and energy loss by electronic excitation. It should be recognized that a significant departure was taken in this analysis to include the effect of thermal neutrons. Although their effect appeared to be noticeable only when thermal neutrons were about an order of magnitude greater than fast neutrons, it was seen that this contribution helped to correlate the mechanical property data.

The work of Sheely clearly points the direction for future studies that may well result in development of a universal damage parameter. In fact, a recent study by McElroy, Dahl, and Serpan [*21*] showed without question that correlation of mechanical property changes from irradiations in significantly different reactors could only be correlated if all neutrons from thermal energy levels on up to 18 MeV were accounted for. This study also suggested that a damage function descriptive of the damaging process was dissimilar for different materials and irradiation temperatures. The damage functions [*21*] are not unlike the damage cross sections of Shure and Rossin but are derived by adjustment of input models by the actual data themselves and include values from 10^{-10} to 18 MeV. Both of these techniques are not at a stage of development yet where all parameters are available for use by the radiation effects community. Accordingly, other techniques that have been shown to be applicable for reactor surveillance damage analysis will be considered here.

The neutron spectral analysis technique used at NRL [*16*, *19*] is an extension of the work of Dahl and Yoshikawa [*22*]. This technique uses the approach of determining the iron activation cross section for a reactor spectrum calculated by a reactor physics computer code. It is then a simple process to convert any fluence value into one which reflects this newly calculated cross section based upon a calculated neutron spectrum. In their analysis, Dahl and Yoshikawa found that accounting for neutrons of energies greater than 0.5 MeV in several different reactor spectra would yield the best overall correlation of irradiation data developed from exposures in those reactors. A recognized limitation of this approach is that it claims the same effect for neutrons of energies of 0.5 MeV as for neutrons of energies of 1, 5, or even 10 MeV. No other correlation point was found by Dahl and Yoshikawa to be as effective in normalizing diverse data points; however, no spectra typical of power reactors were considered. A distinct feature of this technique is that it retains the concept of n/cm^2 as well as the same order of magnitude of fluences.

Big Rock Point Reactor Spectrum Analysis—The application of this technique as used for the BRPR is presented in Table 42*a* for n/cm^2 >0.5 MeV and in Table 42*b* for n/cm^2 >1 MeV; it is summarized for the n/cm^2 >0.5 MeV case as follows. Spectral averaged cross sections, $\bar{\sigma}$, for specific locations in the BRPR are evaluated for n/cm^2 >0.5 MeV using the relationship:

$$\bar{\sigma} = \frac{\int_0^\infty \sigma_{Fe}\phi(E)\, dE}{\int_{0.5}^\infty \phi(E)\, dE} \qquad (3)$$

where σ_{Fe} is the differential cross section for iron as constructed by Helm (listed in Table 41), and $\phi(E)$ is the flux spectrum, also tabulated in Table 41 for the BRPR. The resulting values are listed in Table 42*a* and 42*b* as part of the development of a maximum-power 40-year fluence value for the BRPR vessel wall in terms of n/cm^2 >0.5 MeV and >1 MeV, respectively.

Column 1 gives the location of origin of the activation data.

Column 2 lists the instantaneous flux values for the reactor location during the time of the radiation damage exposure. They are based upon a fission spectrum and a fission-averaged cross section, $\bar{\sigma}_{fs}$, of 68 mb for iron. They also are listed for the as-irradiated reactor power level of 157.9 MWt. For simplicity, an extrapolation of the vessel wall surveillance flux only is presented.

Column 3 shows the flux values first extrapolated to 240 MWt using a simple ratio of power levels—240/159.7. This yields a value of 9.24 × 10^{10} n/cm^2·s^{-1}. The flux is then revised with regard to its fission spectrum cross-section base from 68 mb to Helm's value of 82.6 mb. This change is accomplished using the relationship:

$$\text{flux}_1 \times \text{cross section}_1 = \text{flux}_2 \times \text{cross section}_2 \qquad (4)$$

Great care must be exercised at this point to assure that the flux and cross section on each side of the equation are based on the same criterion (as suggested by the subscripts). To make this conversion for Column 3, we must use cross sections for flux >1 MeV on both sides since this is the criterion for both fluxes. Note that both the fission-spectrum-averaged cross sections of 68 and 82.6 mb are values averaged over *the entire fission*

TABLE 42—*Development of maximum neutron fluence values for 240 MWt operation of the Big Rock Point reactor pressure vessel.*[a]

1	2	3	4	5	6	7
A. Extrapolation Based on Calculated Spectrum, n/cm² > 0.5 MeV						
Flux Monitor Location	Fission Spectrum for $\phi_{fs} > 1$ MeV[b] $\bar{\sigma}_{fs} = 68$ mb (n/cm²·s⁻¹)	$\phi_{fs} > 1$ MeV (n/cm²·s⁻¹)	$\bar{\sigma}_{cs}$ (mb) Spectrum-Averaged Cross Section for $\phi_{cs} > 0.5$ MeV	ϕ_{cs} (>0.5 MeV)[c] Adjustment of Fission-Spectrum (n/cm²·s⁻¹)	Extrapolated ϕ_{cs}[d] at PVW/SS Clad Interface n/cm²·s⁻¹ > 0.5 MeV	Maximum Fluence Φ_{cs} (n/cm²)
Vessel wall.........	6.15×10^{10}	7.61×10^{10}	126.0	7.14×10^{10}	6.3×10^{10} (0.8813)	7.9×10^{19}
B. Extrapolation Based on Calculated Spectrum, n/cm² > 1 MeV						
Flux Monitor Location	Fission Spectrum for $\phi_{fs} > 1$ MeV[b] $\bar{\sigma}_{fs} = 68$ mb (n/cm²·s⁻¹)	$\phi_{fs} > 1$ MeV (n/cm²·s⁻¹)	$\bar{\sigma}_{cs}$ (mb) Spectrum-Averaged Cross Section for $\phi_{cs} > 1$ MeV	$\phi_{cs} > 1$ MeV Adjustment of Fission-Spectrum (n/cm²·s⁻¹)	Extrapolated ϕ_{cs}[e] at PVW/SS Clad Interface n/cm²·s⁻¹ > 1 MeV	Maximum Fluence Φ_{cs} (n/cm²)
Vessel wall.........	6.15×10^{10}	7.61×10^{10}	171.0	3.39×10^{10}	2.76×10^{10} (0.812)	3.5×10^{19}

[a] All data except for Column 2 are extrapolated to 240 MWt operation and $\bar{\sigma} =_{fs}$ 82.6 mb.

[b] For 159.7 MWt at the vessel wall location.

[c] $\phi_{cs} > 0.5 \text{ MeV} \times \bar{\sigma}_{cs} > 0.5 \text{ MeV} = \phi_{fs} > 1 \text{ MeV} \times \bar{\sigma}_{fs} > 1 \text{ MeV}\ \phi_{cs} > 0.5 \text{ MeV} = \dfrac{\phi_{fs} > 1 \text{ MeV (Col. 3)} \times \dfrac{82.6}{0.693}}{\bar{\sigma}_{cs} > 0.5 \text{ MeV (Col. 4)}}$

[d] $$\frac{\phi_{cs} > 0.5 \text{ MeV at reactor vessel wall}}{\phi_{cs} > 0.5 \text{ MeV at surveillance location}} = \frac{8.963 \times 10^{-7}}{1.017 \times 10^{-7}} = 0.8813$$

[e] $$\frac{\phi_{cs} > 1 \text{ MeV at reactor vessel wall}}{\phi_{cs} > 1 \text{ MeV at surveillance location}} = \frac{6.078 \times 10^{-7}}{7.484 \times 10^{-7}} = 0.812$$

spectrum, not just part of the spectrum. In order to obtain cross sections for $n/cm^2 \cdot s^{-1} > 1$ MeV, we divide both fission averaged values by 0.693 which is the fraction of neutrons in a fission spectrum with energies >1 MeV. The cross sections for neutrons in a fission spectrum for energies >1 MeV are then:

$$\sigma_{fs} > 1 \text{ MeV} = 68 \text{ mb}/0.693 = 98.26 \text{ mb} \quad (5a)$$

$$\sigma_{fs} > 1 \text{ MeV} = 82.6 \text{ mb}/0.693 = 119.1 \text{ mb} \quad (5b)$$

Substituting into Eq 4:

$$9.24 \times 10^{10} \times 98.26 \text{ mb} = X \times 10^{10} \times 119.1 \text{ mb} \quad (6)$$

$$7.61 \times 10^{10} = n/cm^2 \cdot sec^{-1} > 1 \text{ MeV for Column 3.}$$

Column 4 lists the spectral averaged cross sections for the two BRPR locations as were determined in Eq 3.

Column 5 gives the flux values after the conversion has been made from a fission spectrum to a calculated spectrum as well as from >1 MeV to >0.5 MeV. Again, the relationship of Eq 4 holds and we can substitute:

$$7.61 \times 10^{10}\ n/cm^2 \cdot s^{-1} > 1 \text{ MeV} \times 119.1 \text{ mb}$$
$$= X \times 10^{10}\ n/cm^2 \cdot s^{-1} > 0.5 \text{ MeV} \times 126.0 \text{ mb} \quad (7)$$

$$7.14 \times 10^{10}\ n/cm^2 \cdot s^{-1} > 0.5 \text{ MeV}$$
$$= \text{calculated spectrum flux.}$$

Column 6, the relative flux level for the BRPR vessel wall surveillance location is extrapolated to the PVW/SSClad interface of the vessel wall. This is quite simply done by establishing a ratio between the pressure vessel wall flux for $n/cm^2 > 0.5$ MeV, tabulated in Table 41, and the flux >0.5 MeV for the vessel wall surveillance location in the reactor. The flux ratios shown in Table 42 indicate the pressure vessel flux is 88.13 percent of the vessel wall surveillance location flux. The flux from Column 5 is multiplied by this factor and thus becomes a flux value extrapolated to the inner edge of the BRPR low-alloy steel vessel wall. Column 7 shows the results of integrating the flux value of Column 6 over a period of 40 years, 1.262×10^9 s.

In reflecting upon the fact that both the >1 MeV and >0.5 MeV calculated spectrum fluences for the BRPR were less than the fission spectrum >1 MeV value, it must be remembered that the calculated spectrum shapes are quite different from the fission spectrum and that the fission spectrum is not a realistic distribution to employ.

It should be recalled that only the vessel wall surveillance-location flux was extrapolated. The principal reason for this is that, being so close to the wall, the small amount of extrapolation would necessarily have a com-

TABLE 43—*Calculated neutron flux (arbitrary units) per energy group for three locations in the Yankee reactor.*

Energy Group Structure			Calculated Relative Neutron Flux[a] ($n/cm^2 \cdot s^{-1}$)		
Group No.	Lethargy (U)	Lower Energy Limit (E_L), eV	Accelerated[b]	Vessel Wall[c]	PVW/SS Clad[d]
1	0.25	7.78×10^6	5.409×10^{-4}	1.006×10^{-5}	7.182×10^{-6}
2	0.50	6.07×10^6	1.682×10^{-3}	3.121×10^{-5}	2.260×10^{-5}
3	0.75	4.72×10^6	2.912×10^{-3}	4.345×10^{-5}	3.132×10^{-5}
4	1.00	3.68×10^6	3.385×10^{-3}	4.211×10^{-5}	3.016×10^{-5}
5	1.25	2.87×10^6	4.183×10^{-3}	4.503×10^{-5}	3.172×10^{-5}
6	1.50	2.23×10^6	6.383×10^{-3}	6.271×10^{-5}	4.357×10^{-5}
7	1.75	1.74×10^6	4.923×10^{-3}	5.354×10^{-5}	3.912×10^{-5}
8	2.00	1.35×10^6	4.777×10^{-3}	5.364×10^{-5}	3.847×10^{-5}
9	2.25	1.05×10^6	3.119×10^{-3}	4.642×10^{-5}	3.446×10^{-5}
10	2.50	8.21×10^5	3.491×10^{-3}	4.850×10^{-5}	3.698×10^{-5}
11	2.75	6.39×10^5	3.944×10^{-3}	5.473×10^{-5}	4.060×10^{-5}
12	3.00	4.98×10^5	3.364×10^{-3}	5.082×10^{-5}	4.194×10^{-5}
13	3.25	3.88×10^5	2.298×10^{-3}	3.456×10^{-5}	2.709×10^{-5}
14	3.50	3.02×10^5	2.840×10^{-3}	4.329×10^{-5}	3.106×10^{-5}
15	3.75	2.35×10^5	2.487×10^{-3}	3.866×10^{-5}	2.784×10^{-5}
16	4.00	1.83×10^5	2.182×10^{-3}	3.514×10^{-5}	2.187×10^{-5}
17	5.00	6.74×10^4	6.817×10^{-3}	1.052×10^{-4}	7.345×10^{-5}
18	9.00	1.23×10^3	1.968×10^{-2}	2.900×10^{-4}	1.991×10^{-4}
19	13.00	2.26×10^1	2.180×10^{-2}	2.927×10^{-4}	2.031×10^{-4}
20	17.00	4.14×10^{-1}	2.323×10^{-2}	2.703×10^{-4}	1.828×10^{-4}
21	20.72	1.0×10^{-4}	8.753×10^{-1}	2.099×10^{-3}	6.802×10^{-4}

ϕ Total	1.000	3.751×10^{-3}	1.845×10^{-3}
ϕ ($E_L > 1.0$ MeV)	3.260×10^{-2}	3.979×10^{-4}	2.860×10^{-4}
ϕ ($E_L > 0.5$ MeV)	4.270×10^{-2}	5.422×10^{-4}	3.981×10^{-4}
ϕ ($E_L > 0.18$ MeV)	5.247×10^{-2}	6.939×10^{-4}	5.060×10^{-4}
ϕ ($E_L > 0.067$ MeV)	5.929×10^{-2}	7.991×10^{-4}	5.794×10^{-4}

[a] The low values for the fluxes are only peculiarities of the computer program output. They are not to be construed as real flux values.
[b] 3.515-in. outside core baffle.
[c] 0.75-in. outside thermal shield.
[d] Interface between low alloy steel wall and stainless steel cladding (clad thickness, 0.110 in.).

mensurately small error. Furthermore, from Fig. 64, it can be seen that the vessel wall surveillance location at the fuel core corners is closest to the actual pressure vessel wall, and hence, the neutron flux should be highest there. For comparison, note that the accelerated locations are opposite the flat face of the core, that the vessel is farther from the core in this location, and consequently, the vessel flux will be lower. In establishing the reactor layout for the one dimensional transport-theory programs calculation, however, it was necessary to utilize an average core diameter which made the corner location not as close to the vessel wall but the flat face of the core nearer to the wall. The intensity drop for calculated flux between the accelerated location, the vessel wall surveillance location, and the vessel wall, therefore, was subject to these approximations and was not wholly accurate. This can be observed and verified by comparing the calculated decrease from accelerated to wall surveillance locations with the measured flux values for these same two positions. The calculated flux decrease for n/cm^2 >0.5 MeV is taken from Table 41 as $1.017 \times 10^{-6}/9.924 \times 10^{-6} = 0.102$, while the decrease from measurements is $7.14 \times 10^{10}/1.20 \times 10^{12} = 0.059$ for one accelerated location and $7.14 \times 10^{10}/1.75 \times 10^{12} = 0.041$ for the other accelerated location. (Note that the 1.20 and 1.75×10^{12} values are not shown elsewhere in this report). It is apparent then that the spectral calculation would overestimate the flux at the vessel wall (not allow for sufficient flux degradation) and present a false, overly conservative picture of the potential lifetime fluence. If now the higher of the two accelerated fluxes were taken that is, the 1.75×10^{12} n/cm^2 >0.5 MeV and were decreased by the average of the above-developed factors (from measurements) of 0.5, it would yield a flux of about 7.7×10^{10}. Reducing this again by the 0.8813 factor (see Table 42*a*), relating the vessel wall surveillance location to the vessel wall itself, would yield a flux value of about 6.8×10^{10} n/cm^2 >0.5 MeV. This compares very favorably with the value extrapolated from the vessel wall surveillance location, 6.3×10^{10} n/cm^2 >0.5 MeV. Without doubt, the most important lesson to be learned here is that neutron spectral calculations cannot always be taken at face value but should be verified carefully with flux detector measurements wherever possible.

Yankee–Rowe Spectrum Analysis—A second pressure vessel lifetime analysis was performed for the Yankee–Rowe reactor [*19*]. Energy group fluxes for the accelerated and vessel wall surveillance locations are tabulated with those for the PVW/SSC in Table 43. The detailed lifetime development for 30 years, based upon calculated neutron spectra, is shown in Table 44*a* for n/cm^2 >0.5 MeV and in Table 44*b* for n/cm^2 >1 MeV. The development scheme matches that for the BRPR; for the Yankee, however, extrapolations from both the accelerated and vessel wall surveillance locations are made.

The reason that both accelerated and vessel wall surveillance location

TABLE 44a—*Development of maximum neutron fluence values >0.5 MeV for 600 MWt operation of the Yankee reactor pressure vessel.*[a]

1	2	3	4	5	6	7
Flux Monitor Location	Fission Spectrum[b] for $\phi_{fs} > 1$ MeV $\bar{\sigma}_{fs} = 68$ mb $(n/cm^2 \cdot s)^{-1}$	$\phi_{fs} > 1$ MeV $(n/cm^2 \cdot s^{-1})$	$\bar{\sigma}_{cs}$ (mb) Spectrum-Averaged Cross Section for $\phi_{cs} > 0.5$ MeV	ϕ_{cs} (>0.5 MeV)[c] Adjustment of Fission-Spectrum Flux $(n/cm^2 \cdot s^{-1})$	Extrapolated ϕ_{cs}[d] $(n/cm^2 \cdot s)$ at PVW/SS Clad Interface $[n/cm^2 \cdot s^{-1}$ (>0.5 MeV)]	Maximum Fluence Φ_{cs} (n/cm^2)
Vessel wall.	2.73×10^{10}	2.67×10^{10}	126.5	2.51×10^{10}	1.84×10^{10}	1.46×10^{19}
Accelerated.	3.25×10^{12}	3.28×10^{12}	114.9	3.40×10^{12}	3.16×10^{10}	2.50×10^{19}
	3.06×10^{12}	3.09×10^{12}	114.9	3.20×10^{12}	2.98×10^{10}	2.36×10^{19}
	2.56×10^{12}	2.58×10^{12}	114.9	2.67×10^{12}	2.48×10^{10}	1.96×10^{19}
	1.69×10^{12}	1.70×10^{12}	114.9	1.76×10^{12}	1.64×10^{10}	1.30×10^{19}

[a] All data except for Column 2 are extrapolated to 600 MWt operations and $\bar{\sigma}_{fs} = 82.6$ mb.

[b] For 504 MWt at the vessel-wall location and 490 MWt for the accelerated location.

[c] $\phi_{cs} > 0.5 \text{ MeV} \times \bar{\sigma}_{cs} > 0.5 \text{ MeV} = \phi_{fs} > 1 \text{ MeV} \times \bar{\sigma}_{fs} > 1 \text{ MeV}$

$$\phi_{cs} > 0.5 \text{ MeV} = \frac{\phi_{fs} > 1 \text{ MeV (Col. 3)} \times \left(\dfrac{82.6}{0.693}\right)}{\bar{\sigma}_{cs} > 0.5 \text{ MeV (Col. 4)}}$$

[d] $\dfrac{\phi_{cs} > 0.5 \text{ MeV at reactor vessel wall}}{\phi_{cs} > 0.5 \text{ MeV at surveillance location}}$

TABLE 44b—*Development of maximum neutron fluence values >1 MeV for 600 MWt operation of the Yankee reactor pressure vessel*[a]

1	2	3	4	5	6	7
Flux Monitor Location	Fission Spectrum[b] for $\phi_f > 1$ MeV $\bar{\sigma}_{fs} = 68$ mb (n/cm²·s⁻¹)	$\phi_f > 1$ MeV (n/cm²·s⁻¹)	σ_{cs} (mb) Spectrum-Averaged Cross Section for $\phi_{cs} > 1$ MeV	ϕ_{cs} (>1 MeV) Adjustment of Fission-Spectrum Flux (n/cm²·s⁻¹)	Extrapolated ϕ_{cs}[d] (n/cm²·s) at PVW/SS Clad Interface [n/cm²·s⁻¹ (>1 MeV)]	Maximum Fluence Φ_{cs} (n/cm²)
Vessel wall.	2.73×10^{10}	2.67×10^{10}	171.7	1.86×10^{10}	1.34×10^{10}	1.06×10^{19}
Accelerated.	3.25×10^{12}	3.28×10^{12}	149.3	2.62×10^{12}	2.30×10^{10}	1.82×10^{19}
	3.06×10^{12}	3.08×10^{12}	149.3	2.46×10^{12}	2.16×10^{10}	1.71×10^{19}
	2.56×10^{12}	2.58×10^{12}	149.3	2.06×10^{12}	1.81×10^{10}	1.44×10^{19}
	1.69×10^{12}	1.70×10^{12}	149.3	1.36×10^{12}	1.19×10^{10}	0.94×10^{18}

[a] All data except for Column 2 are extrapolated to 600 MWt operations and $\bar{\sigma}_f = 82.6$ mb.

[b] For 504 MWt at the vessel-wall location and 490 MWt for the accelerated location.

[c] $\phi_{cs} > 1 \text{ MeV} \times \bar{\sigma}_{cs} > \text{MeV} = \phi_{fs} > 1 \text{ MeV} \times \bar{\sigma}_{fs} > 1 \text{ MeV}$

$$\phi_{cs} > 1 \text{ MeV} = \frac{\phi_{fs} > 1 \text{ MeV (Col. 3)} \times \left(\frac{82.6}{0.693}\right)}{\bar{\sigma}_{cs} > 1 \text{ MeV (Col. 4)}}$$

[d] $\dfrac{\phi_{cs} > 1 \text{ MeV at reactor vessel wall}}{\phi_{cs} > 1 \text{ MeV at surveillance location}}$

fluxes were extrapolated to the vessel wall was to test the accuracy of the computer code. The schematic diagram of the Yankee core, Fig. 74, shows that both types of surveillance capsules were in a direct line from the core to the vessel wall; the computer code would calculate the vessel flux from the exact same nuclear data. If the code were properly handling the neutron interactions from the core to the vessel, the extrapolations to the vessel from both surveillance locations should yield closely duplicating results. Inspection of Column 7 on Table 44*a* and *b* reveals that the extrapolated 30-year fluence based upon the vessel wall surveillance data falls within the range of values determined from the accelerated location capsules. The spread from these accelerated capsules is presumed to have been caused by the capsules being displaced from their intended locations by water flow variations or turbulence. A fair conclusion which can be reached from this then is that the computer code apparently is operating in a realistic manner.

The tabulated spectra for the Yankee reactor in Table 43 provide an independent means to further assess the validity of the neutronics of the computer code. The Yankee extrapolation was from two distinct locations in the reactor to a third location; there was no real means to check the accuracy except that both sets of extrapolated data were in general agreement. If this were coincidence, there would be no way to discover it. Note, however, that the flux at the accelerated location can be extrapolated to the vessel wall surveillance location and that a flux value has been measured at the accelerated location to act as a verifying bit of evidence. From Table 43, the flux >0.5 MeV at the vessel wall surveillance location is 0.0127 of the accelerated location flux. From Table 44*a*, Column 5, the flux at the vessel wall surveillance location is decreased from the accelerated location fluxes over the range of 0.007 to 0.014. Despite the relatively wide range of values, the reasons for which have been discussed adequately, the measured flux decrease from accelerated to vessel wall surveillance locations agrees with the calculated spectrum values for the flux decrease. This is indeed significant and lends additional confidence to the value of the flux extrapolated from these two surveillance locations to the vessel wall itself of the Yankee reactor.

SM-1A—LITR Reactor Spectrum Analyses—A third example of spectral calculation application is that for the Army SM-1A reactor vessel lifetime [*13*]. In this study the only real surveillance data were a flux measurement made adjacent to the reactor vessel wall. This extremely valuable bit of data provided the key to a successful analysis however. A radiation response trend for a duplicate forging of the vessel steel was developed in the low intensity test reactor (LITR), and a fluence was determined for the maximum embrittlement increase permissible for the reactor. This LITR-developed limit fluence was next translated into a calculated-spectrum limit fluence. Then the calculated spectrum neutron fluence generated by the SM–1A in one full-time, full-power year was divided into the LITR limit fluence (for producing the limit embrittlement increase). This then yielded the maximum number of full-time, full-power years permissible for the SM–1A reactor, based upon calculated spectrum considerations. The lifetime for the SM–1A previously had been based upon fission spectrum considerations and life was running out. The calculated spectrum analysis showed that the reactor vessel life could be extended *safely* by 35 percent over that predicted by fission spectrum considerations.

Conclusions

Perhaps the most significant realization of recent years is that dosimetry is of high importance to mechanical property measurements so that activity measurements are made as carefully as possible and converted into fluxes using the very best cross-section data available. A further realization which has followed the first is that neutron spectra are vastly different from each other and also from a fission spectrum so that to quote a fluence value for a critical application in any criterion other than one which takes the influence of neutron spectrum into account is to quote an erroneous value.

Fortunately, at this point in time, measurements of flux and fluence are good enough that future improvements in cross sections will only require a ratio adjustment of these values to make them as accurate as the current technology permits. Detailed research efforts now underway will soon yield results concerning the accuracy of the computer code spectral calculations through the SAND–II code. Radiation damage models are being proposed continually to help relate fluence and radiation damage; theories of the effectiveness of neutrons of all energy groups in creating damage also are being formulated. All these efforts have been and will provide better and better accounting means for neutron dosimetry.

References

[*1*] "Standard Method for Measuring Neutron Flux by Radioactivation Techniques," ASTM E 261-70, *1970 Book of ASTM Standards*, Part 30.

[*2*] Hart, R. G., Bigham, C. B., and Miller, F. C., "Silver 109 as an Epithermal Index Monitor for Use with Cobalt Flux Monitors," CRDC-1083, Atomic Energy of Canada, Ltd., Chalk River, Ont., April 1962.

[*3*] Cabell, M. J. and Wilkins, M., "Routine Use of Silver Cobalt Activity Ratio Method to Determine Epithermal Indices for Large Neutron Dose Irradiation," AERE-R-5122, United Kingdom Atomic Energy Authority, Harwell, Feb. 1966.

[*4*] Cabell, M. J. and Wilkins, M., "A Comparison of Pu-240 and Ag-109 as Epithermal Index Monitors for Long Irradiations," AERE-R-4866, United Kingdom Atomic Energy Authority, Harwell, April 1965.

[*5*] *Neutron Cross Sections*, International Series of Monographs on Nuclear Energy, Hughes, D. J., Pergamon Press, New York, 1957.

[*6*] Carroll, E. E., Jr, and Smith, G. C. "Iron-54 (n, p) Cross Section Measurement," *Nuclear Science and Engineering*, Vol. 22, 1965, pp. 411–415.

[7] Helm, J. W., "High-Temperature Graphite Irradiations: 800 to 1200 Degrees C: Interim Report No. 1," Battelle-Northwest Laboratory Report BNWL-112, Sept. 1965.
[8] Barrall, R. C. and McElroy, W. N. "Neutron Flux Spectra Determination by Foil Activation," Air Force Weapons Laboratory Report AFWL-TR-65-34, Vol. II, Aug. 1965.
[9] Shure, K., "Radiation Damage Exposure and Embrittlement of Reactor Pressure Vessels," WAPD-TM-471, Westinghouse Atomic Power Laboratory, Nov. 1964.
[10] Cranberg, L. et al, "Fission Neutron Spectrum of U^{235}," *Physical Review*, Vol. 103, No. 4, 1 Aug. 1956, pp. 662–670.
[11] Grundl, J., "A Study of Fission-Neutron Spectra with High-Energy Activation Detectors–Part II Fission Spectra," pending publication in *Nuclear Science and Engineering*.
[12] McElroy, W. N. et al, "A Computer-Automated Iterative Method for Neutron Flux Spectra Determinations by Foil Activation," AFWL-TR-67-41, Vols. I, II, III, and IV, Air Force Weapons Laboratory, Sept. 1967.
[13] Bohl, H. et al, "P1MG—A One-Dimensional Multigroup P-1 Code for the IBM 704," WAPD-TM-135, Westinghouse Atomic Power Laboratory, July 1959.
[14] Duane, B. H., "Neutron and Photon Transport Plane-Cylinder-Sphere (GE-ANDP) Program S Variational Optimum Formulation," XDC-9-118, General Electric Company, 1959.
[15] Shure, K., "P-3 Multigroup Calculations of Neutron Attenuation," *Nuclear Science and Engineering*, Vol. 19, 1964, p. 310.
[16] Serpan, C. Z., Jr., and Steele, L. E., "Neutron Spectral Considerations Affecting Projected Estimates of Radiation Embrittlement of the Army SM-1A Reactor Pressure Vessel," NRL Report 6474, Naval Research Laboratory, 30 Sept. 1966.
[17] Moote, F. G., Walthousen, L. D., and Clancy, E. F., "Fast Neutron Flux in a Mock-up of SM-1A Core and Vessel," ACNP-63586, Allis-Chalmers Mfg. Co., Aug. 1963.
[18] Rossin, A. D., "Dosimetry for Radiation Studies," ANL-6826, Argonne National Laboratory, March 1964.
[19] Serpan, C. Z., Jr., and Hawthrone, J. R., "Yankee Reactor Pressure-Vessel Surveillance: Notch Ductility Performance of Vessel Steel and Maximum Service Fluence Determined from Exposure During Cores II, III, and IV," NRL Report 6616, Naval Research Laboratory, 29 Sept. 1967.
[20] Sheely, W. F., "Correlation of Radiation Damage to Steel with Neutron Spectrum," *Nuclear Science and Engineering*, Vol. 29, 1967, pp. 165–175.
[21] McElroy, W. N., Dahl, R. W., Jr., and Serpan, C. Z., Jr., "Damage Functions and Data Correlation," *Transactions*, American Nuclear Society, Vol. 12, No. 1, 1969, pp. 138-139.
[22] Dahl, R. E., Jr., and Yoshikawa, H. H., "Neutron Exposure Correlation for Radiation-Damage Studies," *Nuclear Science and Engineering*, Vol. 21, 1965, pp. 312–318.

CHAPTER 6—SURVEILLANCE PROGRAMS CRITIQUE AND RECOMMENDATIONS

General Comments 239

Army Reactor Surveillance Critique 239

- SM-1A 239
- SM-1 241
- PM-2A 241

Commercial Reactor Surveillance Critique 242

- Big Rock Point (BRPR) 242
- Yankee-Rowe 243
- Dresden 246
- Elk River 246
- Humboldt Bay 247

Features to be Considered in Assuring an Adequate Surveillance Program 247

APPENDIX I—Surveillance Program Recommendations 251

APPENDIX II—Suggested Guidelines for Selecting a Surveillance Specimen Evaluation Laboratory 254

References 259

CHAPTER 6—Surveillance Programs Critique and Recommendations

General Comments

The most important factors of any reactor pressure vessel surveillance program are: (1) it provides for the irradiation of steel from the actual pressure vessel, (2) the irradiation location is as close as physically possible to the vessel wall with the irradiation temperature and neutron spectrum duplicating those for the reactor vessel. These criteria thus provide a very useful measure of the value to be vested in the results of each reactor vessel surveillance program discussed.

Proceeding from the premise that a surveillance program is designed adequately to yield pertinent data, a corollary factor must be that the mechanical property measurements are performed carefully on fully calibrated testing equipment and that neutron dosimetry analyses are made accurately with established cross-section, fission-yield, half-life, etc., values. To be resisted is the temptation to omit seemingly minor details of calculations or calibrations from a surveillance report. For example, the iron reaction cross section and half-life for the surveillance studies conducted for the Dresden and Humboldt Bay reactors were sufficiently different from the values used for these same items in analyses of the other reactors that adjustments were made to effect a more meaningful intercomparison. Had these values not been published on the grounds that they were the "reference values" for that time, and thus were well known, the discrepancy between the values used for the other reactors and these two would not have been uncovered.

Within this framework of minimal surveillance criteria, a critique is presented of the surveillance programs in each reactor discussed in this report. The critical factors for a successful program are presented also with a recommendation of features which should be included.

Army Reactor Surveillance Critique

SM-1A

The surveillance program for this reactor suffered because of the physical limitations imposed by the reactor design and by the program initiation during the very late stages of construction. By far the most serious drawback to the program was the inability to obtain mechanical property change data from vessel wall irradiation locations. The above-core surveillance data are not as useful in vessel analysis at this time as they may be in the future when more is understood about the individual contributions of neutrons of different energy levels.

One significant aspect of the SM-1A surveillance program has not been discussed earlier in this report. It is that one of the two vessel wall flux monitor tubes broke away from the intended location and fell down around the control rod drive mechanisms [*1*][1] causing some minor operational problems. The design of the tubing was at fault as it consisted of many short sections mechanically treaded together and staked inadequately; water-flow turbulence undoubtedly worked the pieces apart. The clear observation to be made is that surveillance capsule holders, brackets, etc., must be fastened securely in the reactor with strong supporting members or redundancy of clamps along the length of any item potentially subject to pendulum action or both. Separation of the tubing posed no significant hazard as the tubing pieces were small enough and flexible enough to move among the control rod drives, but this is illustrative of the design problems to be circumvented.

Another design factor problem could have been more troublesome to SM-1A operation. Several capsules were placed in open-top baskets around the top periphery of the core. When these were to be withdrawn one capsule was missing from its basket. During the annealing operations, it became necessary to probe within the steam generator where the missing capsule (or what remained of it) was found. Again, the capsules were not locked securely in position as it had been confidently felt that they were too heavy to be lifted away by the coolant flow. This obviously was not the case. Fortunately, no discernible harm was done, but these are lessons to be remembered.

On the positive side, the SM-1A program was aided immensely by the flux measurement for the vessel wall. Knowledge of this value at once eliminated a great deal of conservatism otherwise necessary for the estimation of a flux value for the vessel wall. Also of high benefit to the program was the availability of steel judged to be representative of the reactor vessel steel. It was included in the numerous experimental irradiations conducted to establish the reference framework of embrittlement versus neutron fluence for the vessel steel.

Despite the massive experimental efforts accomplished in support of the reactor and the detailed neutron spectral analyses conducted, the pressure vessel condition of the SM-1A can be estimated only because no mechanical property measurement can be made at the vessel wall. Because of this, the

[1] The italic numbers in brackets refer to the list of references appended to this chapter.

vessel condition always will be susceptible to meaningful reanalysis, either favorable or unfavorable, as new means are developed for interpreting radiation effects data from projections from other reactors.

SM-1

The SM-1 reactor surveillance program is quite marginal at best. Neither vessel steel specimens nor neutron flux monitors were exposed at the pressure vessel wall nor in accelerated locations. The above-core surveillance location for the SM-1 reactor suffers from the same drawbacks as noted for the SM-1A. Another serious drawback is that the documentation of preirradiation characteristics of the pressure vessel steel is marginal and that absolutely none of the vessel steel exists upon which studies might have been performed.

The only factor presently to the advantage of the SM-1 is its low operating history. Regardless of the transition temperature increase rate for the SM-1 steel, it probably has not yet risen significantly because the total operating life is limited since the plant is used primarily for operator training. Strong efforts are presently underway to strengthen the knowledge base of the SM-1 in terms of embrittlement versus vessel wall fluence. Most significant is the exposure of the flux monitor wires in a corner fuel element location of the core. The resulting flux data are to be extrapolated therefrom to the vessel wall using a neutron spectrum calculation to effect the translation. A reasonable trend for increasing vessel embrittlement can be drawn from evaluation of reference A212-B steel radiation response and the small amount of preirradiation data available. Unfortunately, these estimates and projections always include a veil of conservatism as to lifetime values for the plant, thus making rehabilitation measures, such as vessel annealing, mandatory sooner than probably would have been necessary if proper surveillance data were available.

PM-2A

There never was any in-reactor surveillance program for the PM-2A reactor. The exceptionally small pressure vessel size effectively frustrated all attempts to place even flux monitor wires near the vessel. This was unfortunate since projection of vessel condition was forced to depend upon conservative estimates to assure safety. This type of analysis was instrumental in the decision to terminally shutdown the reactor on the grounds that the vessel flux was high and that the resulting embrittlement would be sufficient to cause significant concern for the integrity of the vessel.

The postmortem material and dosimetry analyses showed that the vessel flux level was high but was not high enough to produce sufficient embrittlement to validate the early shut down decision on these grounds alone. Examinations of the PM-2A vessel steel showed very good correlation of irradiation response with that for the duplicate forging steel which was used in the test reactor irradiations to define an embrittlement trend for the steel. This correlation was good enough that, had even the vessel flux/fluence been known, it would have indicated a favorable enough condition to preclude early shutdown for reasons of vessel embrittlement.

An important lesson is to be learned from the PM-2A reactor: no surveillance program within a reactor force a too conservative analysis of embrittlement conditions. The lack of a worthwhile program can do that also. Another point to remember is that extremes of flaw, stress, and temperature were required to fracture the PM-2A vessel, thus providing encouragement to the integrity of the commercial reactor vessel though these results are not a signal to minimize surveillance.

Commercial Reactor Surveillance Critique

Big Rock Point (BRPR)

The pressure vessel surveillance program for this reactor is nearly as complete as realistically possible in view of initially projected vessel wall flux levels. Surveillance steel from plates of the actual reactor were used although some heat treatments corresponding to vessel post-welding stress reliefs were only simulated. Weld metal and heat affected zone specimens machined from a fabrication test weldment simulating the full vessel welds were of significant benefit as they revealed the weld metal to be of higher sensitivity to neutron embrittlement than plate or HAZ. Capsules were placed at vessel wall, accelerated, and thermal control locations for the widest possible coverage. Neutron flux monitor wires were included in each capsule to aid in prompt and simple dosimetry determinations; specimen drilling and radiochemical separation was unnecessary to obtain material for specific dosimetry analyses.

Although the surveillance steel was highly representative, the method of taking weld metal Charpy specimens (notches parallel to plate surface) is not considered optimum, and thus some of the value of the material is lost. Inclusion of the HAZ specimens also was considered highly desirable, but in some specimens the *V*-notch root was not placed directly on the fusion line between base and weld metal. This is evident from the photomacrographs of fusion line locations for two BRPR Charpy specimens in Fig. 104. The results of examining many unirradiated weld and HAZ specimens are shown in Fig. 105 which documents the success (and failure) of placing notch roots on HAZ fusion lines.

A standard reference material composition of the reactor vendor was included in the surveillance program, but no other test reactor irradiation effects data for this steel are published so its correlation potential is

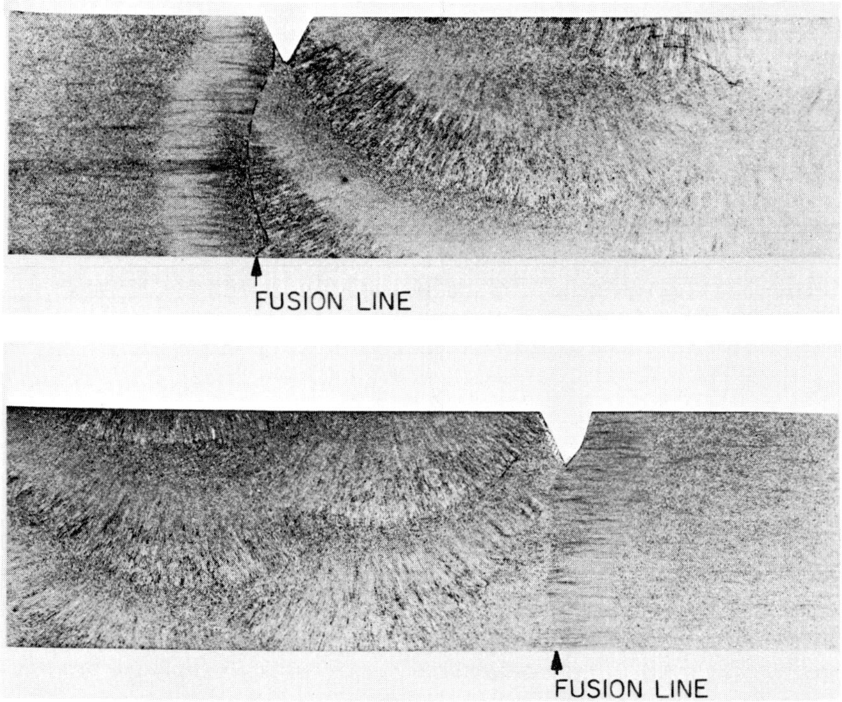

FIG. 104—*Photomacrographs of two Big Rock Point weld HAZ specimens. Dotted line indicates approximate location of fusion line. The notch root of the lower specimen is within Charpy V-notch specifications, although the overall geometry of the notch is not* (*Ref* 1).

low except for General Electric and its customers if applied in a series of surveillance programs. Inclusion of a more generally used reference steel, however, could have yielded enlightening results in view of the overall, apparently low, sensitivity of the BRPR materials.

Although a great deal of documentation was made for the surveillance program, it is unfortunate that a complete chemical analysis was not made and published and that preirradiation mechanical properties were not presented.

Yankee-Rowe

The relatively compact design of the Yankee-Rowe reactor and the lack of planning in design stages for the surveillance program made it difficult to emplace surveillance capsules in appropriate yet accessible locations. Adequate support for these capsules also apparently was a

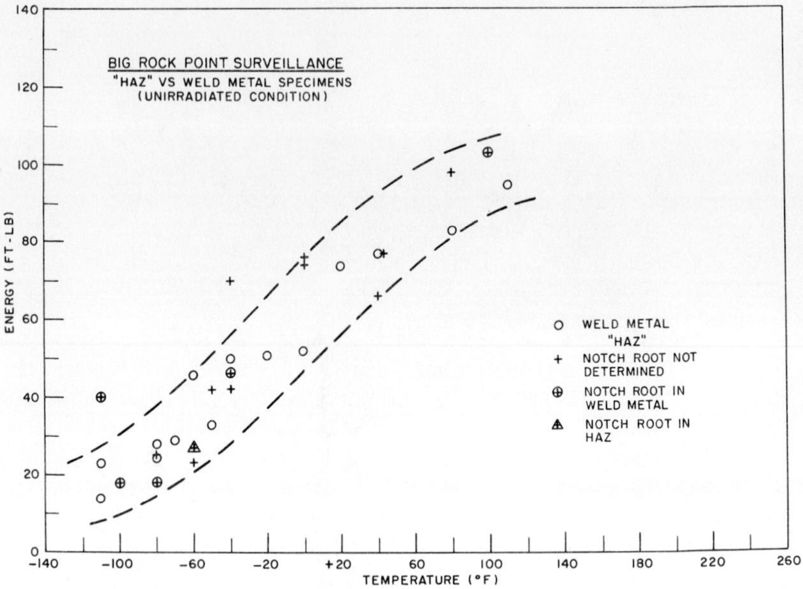

FIG. 105—*Notch ductility characteristics of unirradiated Big Rock Point weld and weld HAZ specimens. Although a similarity of behavior is suggested, the notch root of several HAZ specimens was located in weld metal rather than at the fusion line* (*Ref* 1).

problem as four accelerated location capsules and two more vessel wall location capsules broke off from their mounting fixture and fell inside the vessel. It was postulated [*1*] that the velocity and turbulence of the coolant water flow was such that the capsules acted like pendulums and eventually broke away from their mounts, apparently as a result of fatigue of the capsule supports. Substantial evidence of a pendulum effect was observed on one capsule which did not break away (Fig. 106) [*2*].

Great care was taken in preparing the capsules even to the point of placing shims in the V-notches of the Charpy specimens. However, neutron flux monitor wires were not interspersed in the capsules, nor were any low melting point temperature monitors provided. The absence of flux wires complicated the analysis, requiring the drilling of broken Charpy specimens to permit accurate dosimetry. The absence of temperature monitors was more serious because it thus required an estimation of irradiation temperature based upon calculated gamma heating rates. There is, of course, no guarantee that the temperature monitors will always yield definitive results, but their total absence invariably casts doubts upon the estimated temperature. In the Yankee reactor case, the embrittlement of vessel wall specimens was greater than would be expected for the temperature estimated for the irradiation, thus making the analysis less conclusive than desired.

FIG. 106—*Surveillance capsule sheath showing hole (arrow) caused by fluttering against a fixed reactor member (in-cell photograph) (Ref 2).*

Although the surveillance steel came from the vessel plate, it was heat treated separately with a resulting higher yield strength than the plates in the actual vessel. The surveillance results, therefore, suffer from this fact as well. The absence of weld metal and HAZ specimens, however, is the most serious materials limitation in view of results on the weld in the Big Rock Point program.

A notable positive feature of the Yankee program was the inclusion of specimens of a well-known reference steel. Although thier presence reduced the number of specimens representing the Yankee reactor vessel, they provided valuable confirmation of the embrittlement versus temperature trend and permitted projections far beyond that which might be expected from the limited Yankee program. Further, all elements of the program were well documented and accessible including the important preirradiation mechanical properties.

As the very first commercial pressurized water reactor surveillance program, it proved to be very informative in spite of the shortcomings outlined. The missing weldment specimens and the poor capsule support design are the major negative factors—lessons for the future.

Dresden

The large size of the Dresden boiling water reactor made it relatively easy to emplace the three types of capsules: accelerated, vessel wall, and thermal control. As the first, it lacked important features which later became standard for General Electric reactors. First, neither weld metal nor HAZ specimens were included. The possible implications of this omission have been discussed. The second missing feature related to the lack of a means for temperature monitoring. It was not until the Big Rock Point program that the means of monitoring temperature was devised for General Electric programs.

A notable aspect of the program was the thermal control capsules. These may aid the interpretation of data from the vessel wall specimens in that thermal effects can be separated from neutron radiation effects. This may be important to initial data where aging embrittlement, if any, might represent a significant part of the total. The thermal controls also may be useful in assessing the condition of parts of the vessel which are subject only to temperature cycles. It is, nevertheless, debatable whether the information gained from these specimens for the specific plant is worth the cost. The future use of higher alloy steels which may be prone to thermal metastability, however, may reverse the current questionable attitude on this point.

Two highly unusual exceptions to an otherwise adequate surveillance program must be pointed out. They relate directly to the Charpy specimens. They were taken from representative steel, but from both the top and bottom plate surfaces with notches machined parallel to the plate surface. The effect of the expected superior microstructure in the plate surfaces versus the plate quarter thicknesses (wherein the material characteristics are judged for contractural acceptance), is not well known and always will be the cause for uncertainty in interpreting the Dresden results. To further complicate the results, the V-notches are not cut in the manner considered to be properly representative for characterizing plate steel; hence, the energy absorption upon impact may be somewhat affected by the "non-standard" path of fracture propagation.

Reference steel specimens were not included in the program, but the first surveillance report [*3*] recorded the flux monitor cross-section and half-life values used. This record permitted the values to be adjusted into conformance with more recently obtained values. Accordingly, more meaningful data could be obtained from the reactor and better future projections could be made.

Elk River

By far the two most serious drawbacks to the Elk River Reactor surveillance program are that no material is included that is truly representative of the vessel and that no material surveillance exposures will be obtained at the vessel wall. Because of the far-reaching implications of these two omissions, serious questions exist as to the value of the entire program.

There are, however, certain favorable features to this program. One is the recent effort to measure the flux at the vessel wall and relate it to measurements at the accelerated exposure locations where the current surveillance irradiations are occurring. The employment of many materials provides a wide range of responses so that a reasonably conservative vessel embrittlement projection is not overly difficult. The inclusion of fatigue specimens is useful also, especially since this will be the first data of its kind from a representative power reactor environmental exposure. The effects could be different from test reactor effects. Finally, the inclusion of temperature and neutron flux monitor wires in each capsule was favorable.

Humboldt Bay

The only significant difference between the Humboldt Bay reactor surveillance program and that for the Big Rock Point is the use of only surface thickness Charpy specimens rather than quarter thickness as was done for Big Rock Point. As has been noted, the bias applied to ultimate data interpretations for the reactor based upon this surface thickness material response is not known. A plausible case can be drawn, however, to show that the surface rather than quarter thickness material is more representative because it receives the greatest neutron fluence in a reactor vessel and will be embrittled the most. However, this argument neglects the fact that the notch toughness of the central plate material is generally poorer than the surface layers.

The neutron dosimetry values reported in the surveillance review contained no references to specific cross-section or half-life values. It was presumed that the same values were used for the Humboldt calculations as were published for Dresden. Such uncertainty in the surveillance program results is most important as the embrittlement analysis is based ultimately upon as assumed fluence magnitude which yields uncertainties as to the significance of the measured embrittlement. Thus, great care in data development may be wasted by inadequate knowledge of the exposure.

Features to be Considered in Assuring an Adequate Surveillance Program

Very good guidance is available for planning surveillance programs in ASTM Recommended Practice for Surveillance Tests on Structural Materials in Nuclear Reactors (E 185-70) [*4*]. This document has been drafted by representatives of reactor manufacturers, research organizations, and the Atomic Energy Commission. As with any such group, opinions diverged on certain points, but after full discussion all representatives accepted the contents of the document which was drafted in 1962 and revised slightly in 1966. Guidance for the conduct of mechanical property tests for the irradiated surveillance specimens is given in the ASTM Recommended Practice E 184 [*5*] which supports ASTM E 185. The primary requirements of E 185-70 are presented in outline form in Appendix I. Also, Appendix II, "Suggested Guidelines for Selecting a Surveillance Specimen Evaluation Laboratory," provides important guidance to the proper completion of a surveillance program.

Regarding the materials it must be pointed out that steel from the actual core-region shell courses should be used for surveillance programs and that steel from other shell courses not in the flux field need not be used. The surveillance steel should receive the same heat treatment as the vessel plate, preferably as a portion of the plate. The surveillance weldment should duplicate the fabrication practice for longitudinal seam weld; that is, either a submerged arc or electroslag weld. Postwelding stress relief cycles also should be applied to the surveillance steel even to the extent that it be placed in the furnace with the vessel during stress relief cycles. Exact reproduction of the cold work input to a vessel and the internal and residual stress will probably be impossible to repeat in the surveillance steel. If all the other requirements are met, however, a highly representative material will be used.

Corollary to the qualification of the surveillance material is the necessity to take the specimens from the plate in a meaningul manner. Because plate is qualified at the mill and for ASME Code requirements with quarter-thickness Charpy V-notch specimens, it seems reasonable to use the same material stratum for the surveillance program material. Additionally, the base plate specimens should be taken with orientations longitudinal to the primary plate rolling direction and with notches perpendicular to the plate surface for accurate representation. In addition to these longitudinal specimens, a number of Charpy V-notch and at least one tension specimen of the tranverse orientation also should be included for completeness. While not now an absolute requirement, this could enhance the program and is considered imperative in view of the potential for lowering of the Charpy V-notch shelf by radiation to a point where low energy tearing may take place. Weld metal specimens should be oriented longitudinally to the primary forming direction with notches perpendicular to the plate surface. Orientation of notches at right angles to this will permit the fracture front to run from bead-to-bead in a stop-go manner.

The weld HAZ of a weldment has been discussed adequately previously. It will only be briefly re-emphasized here that insufficient care in preparation will result in specimens representing base metal or weld metal and not the heat affected zone.

The most critical location to a reactor pressure vessel surveillance program is that one nearest to the vessel wall.[2] Ideally, the capsule should be adjacent or against the wall at a point where the fuel is closest to the vessel in order to best represent the peak pressure vessel exposure with distortion caused by variation in neutron energy spectrum. The peak must be sought diligently since, while vessels are round, cores are not, and flux peaking can be expected. However, care must be taken to design the program so no capsule movement occurs.

Thermal control capsules, if used, must be placed in areas of temperature accurately representative of the vessel wall temperature but removed from any significant neutron flux.

The inclusion of neutron flux wires of at least iron as well as bare and cadmium-covered cobalt should be a minimum for all capsules regardless of their intended location or their planned removal date. Removal dates can change for one reason or another, and the dosimetry wires are much more convenient and expedient to handle than drillings from broken specimens. Because the melting point of cadmium is so very close to the operating temperature of most commercial power reactors, it should be encapsulated in some manner or, if necessary, eliminated. In such a case, the non-cadmium covered cobalt wire still should be included. The inclusion of several other wires or foils of dosimeters such as copper, nickel, and titanium can prove to be of high utility as their activation results could be used to aid in refining the neutron spectrum of an irradiation location.

The differences in neutron spectrum between the vessel cladding interface and the surveillance exposure locations may be important. Previous chapters have shown neutron spectral distribution plots for accelerated, vessel wall, and in-wall locations. The principal spectrum effect noted is a hardening of the spectrum; that is, the lower and intermediate energy neutrons are becoming depleted while the higher energy neutrons are not. Therefore, the average neutron energy is rising creating the "hardened spectrum" condition. Research results are now beginning to show that greater radiation damage relative to some threshold fluence is affected by the hardened spectra at power reactor vessel wall environments in contrast to the relatively softer test reactor spectra. The closer the surveillance exposure location is to the pressure vessel wall, the smaller spectral variations become. These even may be negligible. Accordingly, data interpretations and projections from vessel wall capsules are subject to lesser uncertainty bands.

Temperature determinations always will be difficult because thermo-

[2] "Vessel wall" capsules are attached usually to the outer thermal shield and thus usually incur exposures slightly accelerated to the actual vessel. Design of capsule holders whether attached to the thermal shield or other structural member, should assure close and constant proximity to the vessel wall through the life of the vessel.

couple lead access in power reactors is commonly not available. Reliance will thus have to be placed upon the results of inspecting low melting point alloys (see Appendix I for list). These do not give the precise temperature measurement desired, but no other recourse is available. The contribution provided by knowing the temperature is for interrelating the surveillance results with experimental test reactor trends and thereby projecting results to future years. Temperature monitoring is also important to assure that the specimens are exposed at temperatures which truly represent the vessel. This is most important since any encapsulation can only raise the specimen temperatures thus leading to an understatement of the radiation-induced embrittlement as indicated by the fact that a separate trend line could be drawn for the embrittlement of a given heat of steel for irradiations at 50 F increments above 450 F.

Capsule design should be such that the surveillance specimens are not damaged during disassembly and that they can be recovered without undue difficulty. Care must be taken to assure good heat transfer as well for the reasons indicated above.

Withdrawal schedules are based upon the total number of capsules available; sufficient capsules should be withdrawn at one time to permit the comparison of base, weld, and HAZ results for one specific residence period at an irradiation location. At least one vessel wall location capsule should be withdrawn after a year or two in order to measure the fluence to that point. If a dosimeter capsule is available and the fluence is shown to be low, specimen capsule withdrawal is not required. One set of vessel wall capsules nevertheless, should, be withdrawn after about five years. Accelerated location capsules should be withdrawn to yield results in approximately equal intervals with the results representing somewhat greater than the projected fluence. This will aid in accounting for upward revisions in reactor power in later years. If thermal control capsules are in the surveillance program, the initial capsule should not be withdrawn until about five years have passed unless the vessel lifetime fluence is projected to be several times 10^{19} n/cm^2 >1 MeV; in this case early examination of both vessel wall and control specimens would be desirable.

The location of specimen capsules is tied closely to the total number of capsules available. Certainly not less than two and preferably three or four surveillance capsules should be located at the vessel wall for establishment of a trend behavior for base, weld, and HAZ specimens. If accelerated location capsules are to be used for obtaining advance radiation effects information, at least two to three capsules sets should be employed to enable the construction of a meaningful trend. Thermal control capsules yield interesting information concerning the temperature effect. However, they are in no way substitutes for peak flux location capsules. In the face of fiscal or physical limitations, thermal control capsules should be eliminated in favor of vessel wall location capsules, and in no case should either vessel wall or accelerated location capsules be excluded in favor of thermal control capsules.

The surveillance program report should provide an integration of all information that might ever be reasonably desired in the future as reference data. Metallurgical history, melting, rolling, heat treatment, and fabrication procedures as well as *complete* chemical analysis, should be well documented. The later should include an actual iron determination for future reference in case it becomes necessary to use the steel specimens as flux monitor ($^{54}Fe(n, p)\,^{54}Mn$ reaction) materials. Full documentation of preirradiation steel properties is most important. Also, as specimens are taken for surveillance programs, sufficient additional specimens also should be made for establishment of initial material characteristics—and the tests should be made at that time for inclusion in the report. Any remaining surveillance program material (archive material) should be die stamped to identify with the particular reactor and should be stored in a location specifically designated for such controlled, long-term storage. Lack of care in the storage of such valuable pieces of steel have resulted in their loss during general cleanup operations.

The program report is also the proper place to clearly identify the location and orientation of specimens in the parent plate. Items such as the thickness layer within the plate will be apparent. Even such items as the nearness of certain specimens to quenched plate edges can be ascertained; this kind of information can perhaps explain an otherwise anomalous test result. A full statement of nuclear data corresponding to the neutron dosimetry calculations also should be detailed. It is of particular importance to identify the facts that are supported completely as opposed to assumptions or estimates (such as operating temperature determined from a range of temperature monitors).

With careful presurveillance documentation reports, the surveillance testing report can reference many details directly, thus a much more complete (and potentially more valuable) report will result.

APPENDIX I

SURVEILLANCE PROGRAM RECOMMENDATIONS[3]

1. *Materials* (include all at each location)
   (*a*) Base plate (all heats from core region shell course; actual vessel steel).
   (*b*) Weld metal (duplicating core region longitudinal weld).
   (*c*) Weld heat affected zone from weld made with *a* and *b* above.
   (*d*) Reference base metal material (well documented—preferably used in other radiation effects investigations).

[3] This outline based upon authors' experience and review of surveillance data available and upon ASTM E 185-70, "Recommended Practice for Surveillance Tests on Structural Materials in Nuclear Reactors."

2. *Specimens* (include all at each location)
   (*a*) Charpy V-notch (ASTM specifications, number both ends of each) Minimum of eight, preferably twelve Charpy specimens of each material at each location. Transverse Charpy V-notch should be in addition to twelve longitudinal.
   (*b*) Tension (ASTM specifications, number both ends of each) Minimum of two of each material at each location. Transverse tensile should be in addition to two longitudinal.
   (Orientation in plate: longitudinal for primary determination and transverse for full material characteristics).
   (*c*) Others as desired and technology permits (fatigue, fracture mechanics, etc.).

3. *Location*
   (*a*) At vessel wall peak flux points (or at component to be surveyed).
   (*b*) Accelerated exposure locations.
   (*c*) Thermal control locations, out of neutron flux.

4. *Neutron Flux Monitoring*
   (*a*) Flux detectors in each capsule:
      1. Iron.
      2. Cobalt, both bare and shielded with cadmium.
      3. Copper, either cobalt-free or of predetermined cobalt content.
      4. Others (nickel, titanium, scandium; fission foils as technology permits.
   (*b*) Flux measurements at vessel wall or at component to be surveyed. (Early analysis—less than one to two years to be repeated after each major change in reactor operating characteristics which change peak vessel flux.

5. *Temperature Determinations*
   (*a*) Range of low melting point alloys in each capsule. (For typical list see Table 45).
   (*b*) Instrumented determinations if possible.
   (*c*) From operational records (reactor power and coolant temperatures especially helpful).

6. *Capsules*
   (*a*) Corrosion protective sheath, designed and applied for maximum heat transfer.
   (*b*) Baked out just below expected service temperature, or below lowest melting point monitor temperature and seal welded immediately.
   (*c*) Identification number carefully impressed.
   (*d*) Stainless steel or other corrosion resistant materials:
      1. No sheath required.
      2. Unnotched Charpy specimens.
      3. Tensile gage section, oversize.
      4. Fracture mechanics specimens not fatigue cracked.
      5. Flux and temperature monitors secured in stainless steel tubes.

7. *Withdrawal Schedule and Number of Capsules*
   (*a*) Vessel Wall:
      1. Three (minimum) withdrawals of base, weld, and HAZ.

TABLE 45—*Fusible eutectic alloys and metals.*

Melting Point, deg F	Composition, %
430	96.5 Sn—3.5 Ag
464	84 Pb—4 Sn—12 Sb
486	88.8 Pb—11.2 Sb
509	82.5 Cd—17.5 Zn
554	92.5 Cd—7.5 Sb
579	97.5 Pb—2.5 Ag
610	100 cadmium
621	100 lead

2. Initial withdrawal not later than 30 percent of design life. (If initial capsule provides first vessel wall dosimetry, removal at two years or less depending upon dosimeters involved.)
3. Final withdrawal to closely approach end-of-life fluence.

(*b*) Accelerated Location:
1. Two or three (minimum) withdrawals of base, weld, and HAZ.
2. Withdrawals spaced evenly, final to duplicate or exceed end-of-life.

(*c*) Thermal Control:
1. One or two minimum withdrawals of base, weld, and HAZ
2. Initial withdrawal at about 5 years

8. *Program Report*

(*a*) Preirradiation properties data—with exact description of material the specimens represent.
(*b*) Processing, heat treatment, and fabrication history; full chemical analysis—usual elements plus copper, vanadium, alumium, nitrogen, and oxygen.
(*c*) Special description and dimensions.
(*d*) Diagram of specimens located in parent material.
(*e*) Diagram of capsule locations and identification system.
(*f*) Drawing of each capsule showing relative location of all components.
(*g*) Special instructions and features.
(*h*) Details of reactor operating history—power output by month (or less) for exposure period.

9. *Testing Report*

(*a*) Tension and special testing at operating temperature.
(*b*) Testing instrument calibration certification. (See "Suggested Guidelines for Selecting A Surveillance Specimen Evaluation Laboratory").
(*c*) Test results(mechanical properties determined), neutron flux and fluence assumptions, cross sections, half lives, exposure temperature, reactor operating history, etc.
(*d*) Special features or observations.

APPENDIX II

SUGGESTED GUIDELINES FOR SELECTING A SURVEILLANCE SPECIMEN EVALUATION LABORATORY

Background

In order to assess the effects of nuclear radiation upon the properties of the steel of a reactor pressure vessel, specimens of the vessel steel are exposed just inside the vessel, removed periodically, and tested for assessment of radiation induced changes. Such programs of surveillance are required by the U. S. Atomic Energy Commission since the extent of changes thus are determined and used to project the potential changes to the vessel over its lifetime or some significant portion of its lifetime.

Nuclear radiation and its attendant potential for detrimentally changing the properties of the containment vessel steel was new to the many utilities who own nuclear power plants. Consequently, there were strong expressions of the need for guidance relative to aid with surveillance programs. This report outlines several important factors for selecting a laboratory to conduct the periodic vessel surveillance specimen evaluation. It is intended that this report though intended primarily for reactor owners, will be of value to organizations who design and construct reactors and to the prospective organizations who would aid the reactor owners through vessel surveillance analysis.

Introduction

Criteria for the selection of a laboratory to perform postirradiation analyses of reactor surveillance specimens are complicated by the radioactivity of specimens removed. The guidelines delineated here assume that a satisfactory surveillance program has been established for the reactor and that it meets, as a minimum, the requirements set forth in ASTM Recommended Practice for Surveillance Testing, E 185. It is further assumed that a sufficient number of carefully encapsulated steel specimens have been placed in the reactor and that a full description of the initial steel properties and of the surveillance capsules has been provided. The guidelines generally assume a capability for testing, data presentations, analysis, and evaluation. It should be noted at the outset, however, that the presentation of good postirradiation data may suffice if the reactor owner has an in-house (or consultant or subcontractor) capability for evaluation and safety analysis.

An abbreviated sample questionnaire suggesting some minimum requirements of a prospective contractor laboratory is attached. The listing is of specific items required for safe handling and meaningful evaluation of reactor pressure vessel surveillance specimens. (An amplified description follows each item.)

Primary Requirements

1. *AEC License for Possession and Handling of Radioactive Byproduct Material*—A primary requirement for a laboratory to be engaged in the evaluation of irradiated structural material test specimens is that it have an AEC or Agreement State license for radioactive (byproduct) material.

Regulations require that the reactor operator ship irradiated specimens only to properly authorized laboratories. Before the irradiated specimen is shipped it is good practice for the reactor operator to request from the laboratory a signed statement similar to the following:

The ________________ laboratory is properly authorized by virture of (*AEC*) (State) License No. ________ to receive radioactive material in irradiated test specimens from the ____________ ____________ (reactor operator) company.

Signed, ________________, Date__________

A license application can be obtained from the AEC or Agreement State as appropriate. Issuance of the license is evidence that the laboratory has personnel trained and experienced in radiation protection, radiation measurement and monitoring, and in the biological effects of radiation. The license also indicates that the laboratory has some type of radiation handling facilities, a waste disposal capability, and, probably more importantly, that it is an established laboratory of a responsible company.

The types, forms, and end uses for radioisotopes are so varied, each requiring a Byproduct Materials License that license possession obviously does not guarantee the capability for pressure vessel surveillance specimen evaluation. Examination of the license along with the other factors listed below will assist the reactor operator in evaluating the capability of the laboratory to handle reactor surveillance program materials.

2. *β–γ Radiation Shielded Facilities and Remote Handling Capabilities*—The irradiated surveillance specimens returned from the power reactor usually will be encased in radioactive stainless steel sheaths. These radiation levels will doubtless require facilities and experience in the handling of such materials. However, the remote handling of radioactive specimens is the responsibility of the chosen laboratory. Nevertheless, the reactor operator (or surveillance sponsor) may wish to know of the available hot cell capabilities or other shielded testing facilities.

Although possession of proper radiation handling facilities is not a measure of the laboratory's ability, it does add confidence to anticipated ability.

3. *Capsule Breakdown Capability*—The initial concern in specimen evaluation is the careful disassembly of the surveillance capsules and the orderly recovery of their contents: specimens, neutron flux monitors, and eutectic temperature monitors, if included. This requires a radiation handling machine shop-type capability. Makeshift arrangements may not be satisfactory since, for example, bending or notching of tension specimens can invalidate the tests.

4. *Mechanical Property Measurements and Experience in Mechanical Property Evaluation*—

(*a*) *Notch ductility*—Of the specimens usually contained in reactor pressure vessel surveillance programs, Charpy V-notch impact specimens are of particular importance in defining vessel embrittlement and for indicating the potential for brittle fracture. Thus, the laboratory must have a suitably shielded impact test machine which has been certified as accurate with standard Watertown Arsenal Charpy V-notch specimens. The impact testing must be in accordance with the ASTM Standard Method for Notched Bar Impact Testing of Metallic Materials.

(*b*) *Tension tests*—Tension test specimens are common to virtually every surveillance program. One of the principle benefits gained from their analysis is data concerning the ability of the steel to sustain plastic deformation without abrupt failure. This is indicated vividly by the ratio of yield to ultimate tensile strength which, with prolonged (very severe) irradiation, can approach unity, indicating little or no ability for plastic deformation before failure. The availability of sufficient ductility after irradiation, of coarse, is indicated by significant levels of uniform elongation and by relatively large percentages of area reduction during testing. Since the most meaningful data are those obtained at reactor operating temperatures, the prospective laboratory should be capable of performing tension tests at the reactor operating temperature and of providing accurate data on measured uniform and total elongation and reduction of area in addition to 0.2 percent yield and tensile strengths. The tension testing must be in accordance with the ASTM Standard Method for Tension Testing of Metallic Materials, E 8.

(*c*) *Other fracture tests*—If other types of specimens (for example, fracture mechanics) are to be analyzed from a surveillance program, the reactor owner or surveillance sponsor should assure himself that the evaluating laboratory has the capability (preferably by reputation) to test such specimens and properly analyze the results. (Other laboratory test capabilities should be listed; Rockwell hardness testing is considered important enough to be a requirement for the laboratory since it can provide a quick and easy indication of radiation hardening and embrittlement.)

(*d*) *Experience in mechanical property evaluation*—From all aspects of specimen evaluation, the reactor owner must assure himself that the laboratory has the facilities and capability to perform the task satisfactorily. By far the best recommendation is that presented by the reports produced for previous evaluations or from internal or contractural research programs. A laboratory with no experience or a marginal record should be considered very critically since one surrenders to them surveillance specimens which cannot be replaced. The information to be gained is so important, and ultimately so valuable (or costly), that bids should be considered in terms of the bidder's qualifications rather than in absolute dollar values. As a specific example, linear elastic fracture mechanics evaluations are complex, and this complexity is accentuated by radiation. Further, interpretation of fracture mechanics data relative to the more conventional tests is most important. This requires an experienced metallurgical laboratory. Nevertheless, data can be obtained by a testing laboratory and evaluated by another organization or consultant.

Secondary requirements

The following are necessary to a meaningful surveillance program but are not absolute requirements of the prime laboratory since they can be purchased by subcontract or provided by the reactor owner.

5. *Experience in Data Interpretation Relative to Safety*—If a laboratory has a sound capability in analyzing mechanical properties of irradiated materials, the safety implications will be clear and a meaningful presentation should be obtained. Previous experience in such matters need not be a requirement, however, since a *capable* laboratory should be able to produce a satisfactory analysis. Alternatively, data properly presented to the reactor operator or his consultant on safety could discharge the laboratory's obligation.

6. *Neutron Dosimetry and Spectrum Analysis*—Any laboratory desirous of performing surveillance specimen analyses must have an equally competent capability in neutron dosimetry. This may, however, be done by subcontract provided the subcontractor has a sufficiently strong capability. The importance of neutron dosimetry cannot be underestimated: mechanical property changes lose their significance without corresponding dosimetry values; hence, any projections of vessel properties at some future date are dependent upon accurate dosimetry.

In this particular case, it strongly is recommended that the dosimetry laboratory have participated in a "round-robin" type intercalibration. This is an excellent way to uncover unrecognized laboratory biases, and to assure that the results are otherwise consistent with those of other laboratories.

The recent intensified interest in neutron spectral analysis shown by licensing and regulation activities dictates that the surveillance analysis laboratory either possess or be in a position to contract for a strong, capable reactor physics group to perform these analyses.

Further, it is then mandatory that the spectral data be interpreted and applied correctly. Experience in comprehensive analyses combining mechanical property changes as a function of calculated spectrum neutron fluence, extrapolated to service lifetimes, is considered highly desirable but cannot be defined as mandatory.

7. *Waste Disposal Capability*—The reactor owner will wish to assure himself that, upon completion of the destructive evaluation portion of the specimen analysis, the waste specimens and irradiation capsules will be disposed of properly by the contracting laboratory and not be returned to him.

8. *Shipping Containers*—The reactor owner will want to know if the laboratory has its own shielded shipping container(s) or if consideration should be made in the bid evaluation for the supplying of shipping containers by the reactor owner. The laboratory should be requested to certify the radiation limits of their carriers as detailed under Department of Transportation (DOT) Specification 55 (formerly ICC-55), and possibly to supply copies of the carrier drawings which served to acquire the DOT permit. It is important to recognize that the old American Association of Railroads, Bureau of Explosives (BE) permit numbers are no longer valid since the DOT has assumed these functions. The laboratory may have authority under DOT-55 for shipment of anything up to 200 curies; if the reactor owner recognizes that a particular shipment will be in excess of 200 curies, application by the laboratory for special "one-time" DOT approval will be necessary.

SUGGESTIONS FOR A QUESTIONNAIRE FOR A PROSPECTIVE SURVEILLANCE SPECIMEN EVALUATION LABORATORY

1. AEC Byproduct Material License No. ____________
   Expiration Date ____________
   Name and Address of Laboratory ____________
   ____
   Limits of Byproduct Materials (quantity of various isotopes, etc.) ____________
   ____

2. β–γ Radiation Shielded Remote Handling Facilities. YES NO
   Type ____________
   Description ____________
   Location ____________
   Capacity ____________

3. Capsule Disassembly Capability. YES NO
   Describe ____________

4. Mechanical Property Measurement Equipment:

a. Notch Impact Tester and Temperature Conditioning? YES NO
Describe________________________________
Watertown Arsenal Specimen Calibration? YES NO
Date________________________________
Procedures in compliance with ASTM E 23? YES NO

b. Tensile Tester and Temperature Conditioning? YES NO
Describe________________________________
Limits________________________________
Parameters Obtainable________________________________
Type of Extensometer________________________________
Procedures in compliance with ASTM E 8? YES NO

c. Other Property Measurement Equipment? YES NO
Describe________________________________
Hardness Tester? YES NO

d. Brief description of Mechanical Property Evaluation Experience.
References: ________________________________
____

(As indicated above not all items, especially items 5–8, are mandatory but such experience is desirable and the answers could be indicative of the depth of experience.)

5. Experience in Data Interpretation Relative to Safety. YES NO
Describe. ________________________________
(If answer is "No," what organization do you propose as subcontractor?) (Reactor owner may feel qualified to make this interpretation.)

6. Neutron Dosimetry Analysis Capability. YES NO*
Describe________________________________
Round-Robin Intercalibration. YES NO
Describe________________________________
Neutron Spectrum Analysis Capability. YES NO*
Describe________________________________
Experience in Spectral Analysis Application YES NO*
Describe________________________________
References: ________________________________
____

7. Waste Disposal Capability. YES NO*

8. Possession of Shipping Containers. YES NO*
Describe________________________________
DOT Permit Number________________________________

*If answer is "No," how or by what organization do you propose this phase to be conducted?

References

[1] Serpan, C. Z., Jr., Steele, L. E., and Hawthorne, J. R., "Radiation Damage Surveillance of Power Reactor Pressure Vessels," NRL Report 6349, 31 Jan. 1966, Naval Research Laboratory; also *Transactions*, American Society of Mechanical Engineers, Vol. 89, Series D, No. 1, March 1967, pp. 221–232.

[2] Serpan, C. Z., Jr., et al, "Yankee Reactor Pressure Vessel Surveillance: Evaluation of Specimens Exposed During the Second Core," NRL Report 6179, Naval Research Laboratory, 24 Nov. 1964.

[3] Brandt, F. A. and Alexander, A. J., "Effect of Neutron Irradiation at 550 F on Charpy Impact Properties of ASTM A302 Grade B Steel," *Radiation Effects on Metals and Neutron Dosimetry, ASTM STP 341*, American Society for Testing and Materials, 1963, pp. 212–232.

[4] "Recommended Practice for Surveillance Tests on Structural Materials in Nuclear Reactors," ASTM E 185–70, *1970 Book of ASTM Standards*, Part 31.

[5] "Recommended Practice for Effects on High-Energy Radiation on the Mechanical Properties of Metallic Materials," ASTM E 184-62 (1968), *1970 Book of ASTM Standards*, Part 31.

Glossary

A

austenitize

The treatment of steel wherein it is raised in temperature until all the iron has transformed into austenite.

B

base metal

In steel pressure vessels, the material composition forming the main vessel plate.

beltline

The horizontal plane through a fuel core and extending out to a pressure vessel wall having the highest neutron flux.

C

cadmium ratio

The ratio of the number of neutrons of thermal and epithermal energies detected by a bare cobalt foil to the number of neutrons of epithermal energies detected by a cadmium-covered cobalt foil. Note that this yields a thermal neutron count for the ≥ 0.414 eV level rather than the 0.0253 eV level.

Charpy V-notch specimen

A test specimen for measuring the amount of energy absorbed during fracture upon impact with a standard moving weight. The specimen is 0.394 in. square by 2.165 in. long, having a 45-deg notch cut across one surface, the notch root being 0.001 in. radius; it is described by ASTM Standard Method for Notched Bar Impact Testing for Metallic Materials (E 23-66).

D

dynamic tear test

A test similar to the Charpy V-notch but incorporating the advantages of larger size (minimum used is 7 in. long by 1⅝ in. wide by ⅝ in. thick), more realistic notch (pressed notch or brittle weld), and greater crack run below the notch.

E

electroslag welding

A weld joining process wherein two plates to be joined are set vertically about 3 in. apart. Heavy, water cooled shoes cover the gap between plates and act as dams to hold molten weld filler metal used to fuse the plates together. The shoes also conduct electric current to heat and melt the filler wire as it feeds into the molten pool. Fluxing agents float on top of the molten pool to prevent continuous oxidation. Because of the large grain size inherent to the cooled weld joint, electroslag weldments must be quenched and tempered after welding.

epithermal neutrons

Neutrons of energies greater than thermal and below those of fast neutrons. Also referred to as resonance neutrons because of the many resonance peaks observed for the interactions in these energy levels.

F

fast neutrons

An arbitrary designation for neutrons of energies usually greater than 0.1 MeV. More correctly, they are those above 0.1 MeV for which individual resonances are not resolved in "typical" nuclei.

fission foil

A flux monitor that will undergo fission in the neutron flux field under consideration; fission products such as ^{140}Ba and ^{137}Cs are typical of the isotopes analyzed from the fission to yield flux values.

fission spectrum

A specific distribution of neutrons resulting from fission of ^{235}U. This spectrum is a plot of neutron populations within discrete energy levels or bands. At present, forms of the fission spectrum have been developed by Watt, Frye, Cranberg et al, and Grundl; all of these forms disagree somewhat with the fission spectrum described by multiple foil activation techniques.

flux monitor (foil)

A material (usually a metal) that has been placed in a neutron flux field for activation that can subsequently be measured and translated into an accurate neutron flux value. (Synonyms: flux detectors, foils, flux wires, detectors, dosimeters).

fracture toughness

Usually refers to K_{IC} or a measure of stress to denote inherent resistance to intensity for onset of unstable fracture under linear elastic-plane strain conditions.

H

half-life

A specific length of time during which a specific radioisotope will decay to an activity level half that which existed at the start of the time period.

hard spectrum

A neutron spectrum wherein the average energy of the neutrons is relatively high (fast).

heat affected zone (HAZ)

A fusion region adjacent to the interface between weld metal and base metal of a weldment. (Its microstructure may be altered by the high heat input of the welding process.)

heat treatment

A series of operations involving heating and cooling at specified rates in order to develop specific mechanical properties and microstructure in steels.

L

longitudinal

In rolled steel plate, the plate dimension parallel to the primary rolling direction.

N

neutron cross section

That property of an element describing its relative probability for interaction with an impinging neutron. It is measured in units of area, usually the barn, being equal to 10^{-24} cm; for convenience, the cross section of an element nucleus is just the target area presented by that nucleus to an approaching neutron. The target area or cross section is measured as area on a plane normal to the motion of the neutron and can be considered, in a simple sense, as the area of projection of the actual nucleus on the plane.

neutron dosimetry

The measurement of the neutrons present and in motion through a specific area in a unit of time.

neutron fluence

The neutron flux integrated over the time period of irradiation. This term recently was referred to as the dose or exposure or simply nvt. The International Committee for Radiological Units (ICRU) has recommended that neutron fluence be used exclusively as a descriptor for these other terms.

neutron flux

The product of the neutron density (n/cm^3) and velocity (cm/s); this translates to $n/cm^2 \cdot s$.

neutron spectrum

The distribution of neutrons at a specific reactor location indicating the population existing at discrete energy levels or within bands of specified upper and lower energy limits.

normalizing

A strengthening operation in steel wherein the material is heated until all the iron has transformed into austenite, then removed from the furnace and cooled in air.

notch ductility

The ability of a material that contains a notch-type flaw to deform plastically prior to failure thus absorbing a large amount of the energy input.

P

primary rolling direction

The dimension of a plate that has received the greatest amount of extension due to passes beneath the rolls.

Q

quench

Extremely rapid cooling of a steel plate which usually has just been removed from an austenitizing treatment furnace; it is effected by either spraying with water or total submergence in water.

R

rolling direction

In the reduction of steel ingots into plate, the orientation of the steels with respect to its traverse beneath the rolls.

S

shear

A term indicative of ductility. Full shear energy absorption is the amount of energy absorbed by a notched bar impact specimen when the fracture is ductile,

shelf

For the upper shelf, the maximum energy that is absorbed by a notched bar impact specimen under test.

shell

A cylindrical section of pressure vessel wall used in conjunction with one or more others when all welded together form a complete pressure vessel.

soft spectrum

A neutron spectrum wherein the average energy of the neutrons is relatively low.

stress relief

A heating treatment following tempering and usually after significant mechanical working or welding. The steel is raised in temperature to a level below the tempering temperature for several hours to eliminate residual stresses.

subarc welding

A weld joining process wherein a mechanically driven electrode passes through a joint with the filler wire being fed into the arc which is under a constant blanket of flux agent. The fusing arc is thus "submerged" in flux.

T

temper

A heating treatment following quenching or normalizing wherein the steel is raised in temperature to a level below the austenitizing temperature for several hours in order to impart ductility. Tempering has the effect of simultaneously lowering strength while increasing ductility; it also reduces residual stresses.

thermal neutrons

Neutrons of velocities such that they are in thermal equilibrium with their environment. At 20 C, thermal neutrons of 0.0253 eV have a velocity of 2200 m/s. Measurements for most experimental irradiation studies using bare and cadmium-shielded cobalt yield values for "thermal" neutrons of energies ≤ 0.41 eV. This is the highest energy level of neutrons that cadmium can effectively shield out.

threshold foil

A flux monitor that is activated only by neutrons equal to or greater than a certain energy level or threshold. Neutrons of energies below the threshold cannot interact to a significant degree.

transverse

In rolled steel plate, the plate dimension perpendicular to the primary rolling direction.

Index

A

Accelerated irradiation, 50–58, 145, 158, 178, 192, 201, 204, 205
Annealing, 61–64, 106, 200, 240
  Big Rock Point, 147
  PM-2A, 129
  SM-1A, 113–120
  Yankee, 170
Army package power reactor, 105
ASME code, 21, 26, 248

B

Base plate, 74–79, 248, 251
Beltline flux/fluence
  Big Rock Point, 155, 156, 226–230
  Dresden, 185–188
  Elk River Reactor, 197
  Humboldt Bay, 204
  SM-1, 122
  SM-1A, 109, 114
  PM-2A, 125
  Table, 140–142
  Yankee, 167–170, 174, 230–233
Brittle fracture, 24, 28, 33, 126

C

Charpy V-notch, 21–25, 27, 37, 50–52, 108, 121, 127, 145, 158, 175, 179, 191, 201, 242, 252
Chemical composition, 31, 49, 67, 72, 74–89, 158, 253
  Effects of-
    copper, 6, 74–79, 80–89
    phosphorus, 6, 80–89
    sulfur, 83–89
  Reactor steel and welds-
    Big Rock Point, 144
    Dresden, 175
    Elk River reactor, 190
    Humboldt Bay, 201
    La Crosse, 75
    PM-2A, 107
    SM-1, 107
    SM-1A, 107
    Yankee, 157
Computer code (*see* Neutron spectrum)
Cross rolling (*see also* Orientation), 50
Cross sections, 210, 212–218, 221, 222, 225–227, 237, 232, 234, 253
  Damage, 223, 224
  Iron, $^{54}Fe(n, p)$, 201, 211, 214, 216, 221
  Dresden, 183, 185
  Libraries, 129, 169, 219

D

Damage model, 223–225
Ductile-brittle transition temperature (*see* Transition temperature)
Detector (monitor) flux, 108, 109, 145, 153, 183, 185, 192, 209–217, 249, 252
Dose (flux) rate, 56–58
  Yankee, 165
Drop-weight test, 23–26, 37
Ductile fracture, 21, 33, 37–41
Dynamic tear test, 37–45

E

Embrittlement, 19–23, Chapter 2, 139–142
  Neutron in-
    Big Rock Point, 145–158
    Dresden, 174–183, 188
    Elk River, 193–197, 200
    Humboldt Bay, 202, 204
    PM-2A, 124–134
    SM-1, 122–123
    SM-1A, 112–120
    Yankee, 159–170, 174
  Thermal, 183, 189, 246
  Through thickness, 128, 134, 170, 172, 173, 179, 188
Embrittlement trend, 52–55, 58, 59, 66, 67, 70–72, 105
  Elevated temperature, 52, 53, 55, 58, 59
  Reactors-
    Big Rock Point, 152, 155
    Dresden, 180, 182, 186–188
    Elk River, 196–200
    Humboldt Bay, 202, 205
    PM-2A, 128, 133, 242
    SM-1, 123, 241
    SM-1A, 110–114, 240
    Yankee, 53, 163, 169, 172, 173

F

Failures, 26, 27, 29, 33, 158, 240, 241, 244, 245
Fatigue, 32, 35, 97
  Elk River, 198–200
  PM-2A, 98, 125
Fission foil (*see* Detector)
Fission spectrum (*see* Neutron spectrum)
Flaws, 23–25, 26, 29–30, 32–37, 43
  Artificial, 124–126
Forging
  Elk River, 189–191
  PM-2A, 124, 127, 129, 130, 134
  SM-1A, 107, 112, 113, 117, 118, 120
Fracture analysis diagram, 26, 27
Fracture mechanics, 31, 32, 33–37, 131–134, 256
  PM-2A, 34, 35, 124–126, 131, 132

H

Heat affected zone, 31, 49, 74, 76, 140, 141, 143, 145, 147–152, 156, 158, 175, 179, 181–184, 201–263, 204, 243–246, 248, 250, 251, 253
Heat treatment
  Reactors-
    Big Rock Point, 144
    Dresden, 175
    Elk River, 189–191
    Humboldt Bay, 201
    PM-2A, 107
    SM-1, 107, 121
    SM-1A, 107
    Yankee, 157
  Stress relief annealing, 248, 251, 253
Heavy section steel technology program, 30–33
Hydrogen embrittlement, 65–67, 99

I

Industry cooperative program, 30
Industry funded program, 30
Inspection, 29–33
Izod specimens, 121

L

Longitudinal (*see* Orientation)
Low energy tearing, 22, 23, 25, 51, 52, 79, 80, 179

M

Microstructure, 31, 49, 67, 71–74, 79–82, 164, 165
Monitor, flux (*see* Detector)

N

Nil ductility transition temperature (NDT), 22–27, 32–34, 39, 42–44, 127, 128, 139, 140–143, 145, 172, 173, 179, 197
Neutrons
  Thermal, 211, 214–217, 224
  Fast, 109, 183, 214
Neutron dosimetry (flux monitoring), Chapter 5, 28, 120–122, 124, 139, 143, 153, 154, 163, 168, 172, 173, 183–188, 192, 201, 239, 249, 252
Neutron fluence (*see also* Beltline flux/fluence), 49–55, 124, 139–145, 155, 156, 167–169, 180–188, 194–197, 204, 249
Neutron flux (*see also* Beltline flux/fluence), 56–58, 109, 155, 172, 173, 249, 252
Neutron spectrum, Chapter 5, 53, 105, 109, 133, 153–156, 167–170, 183
  Calculations, 109, 125, 133, 167, 168, 217, 232, 241
  Computer codes, 125, 167, 168, 217, 232, 241
  Fission spectrum, 133, 134, 154, 217, 232
  Graphic representations, 220–223
    Big Rock Point, 153, 154
    SM-1A, 110, 113, 219, 220
    PM-2A, 128
    Yankee, 167, 168
    Watt fission, 153, 167, 214, 215, 219, 220
Nozzle cutout, 189

O

Orientation, 21, 50, 51, 204
  of specimens and V-notch, 21, 50, 124, 146–151, 175, 202, 204, 242, 243, 246, 248, 252
  Primary rolling direction, 21

P

Pressure vessel, pressure vessel steel, 140–142
  Big Rock Point, 144, 145
  Dresden, 175
  Elk River, 189–191
  Humboldt Bay, 201
  PM-2A, 107, 122, 123
  SM-1, 107, 241
  SM-1A, 107, 112, 113, 117, 240
  Yankee, 157, 158
Pressure vessel research committee, 29, 30

R

Reference steel, 51, 251
  ASTM A212-B, 57, 189–191
  ASTM A302-B, 51, 52, 54, 58, 59, 63, 70, 73, 78, 139–143, 160–174, 245
  General Electric, 140, 141, 204, 242, 243
Rolling direction (*see* Orientation), 50

S

Saturation, embrittlement, 52, 54, 194
Sensitivity, embrittlement, 49, 55, 67–69, 70–89, 99, 106, 120, 155–158, 163–165
Shear energy absorption, 21–25, 37–39, 50–52, 74–77, 182, 183, 202
Shelf (*see* Shear energy absorption)
Shell course, 175, 200, 201, 248, 251

Shift (*see* NDT Increase and Transition temperature increase)
Spectrum (*see* Neutron spectrum)
Stresses, 23, 25–27, 31, 32, 34–37, 124–127, 131–134
  Experimental irradiation, 65
Stress corrosion cracking, 35, 36, 125
Surveillance capsules, 143, 249, 250, 252, 255
  Big Rock Point, 145, 146
  Dresden, 175, 176
  Elk River, 191, 192
  Humboldt Bay, 201
  SM-1A, 108, 115, 240
  Yankee, 158, 159, 244, 245
Surveillance programs, 19, 20, 105, 106, 139–145, 239
  Reactors-
    Big Rock Point, 145, 242
    Dresden, 175, 246
    Elk River, 189, 246
    Humboldt Bay, 200, 247
    PM-2A, 123, 241
    SM-1, 120, 241
    SM-1A, 107, 239
    Yankee, 158, 243
  Guidelines, 254–258
  Recommendations, 139, 247–253

T

Temperature, irradiation, 49, 57–61, 99, 106, 156, 165, 166, 174, 194, 197, 249, 250, 252, 253
Temperature monitor, 143, 145, 174, 201, 244, 246, 250, 252, 253
Tensile properties, 56, 57, 90–96, 127–131, 183, 197, 203, 252, 253, 255
Thermal control, 179, 183, 201, 246, 249, 253
Thickness effects, 42–45
Transition temperature (*see* Transition temperature increase)
  Curve (illustrative), 21, 22, 50, 51
Transition temperature increase (*see also* Embrittlement trend), 21, 22, 50, 51
  Service limit projection, 139–145
    Big Rock Point, 153–156
    Dresden, 183, 187, 188
    Elk River Reactor, 196, 197
    Humboldt Bay, 204, 205
    PM-2A, 125, 128, 129
    SM-1, 122, 123
    SM-1A, 113, 114, 119
    Yankee, 167–170
Transverse (*see* Orientation)

W

Weld metal, 29, 74, 77–80, 88–91, 145–149, 152, 155, 156, 175, 179, 181, 182, 184, 188, 201–203, 246, 248, 250, 251, 253
  Submerged arc, 78, 248
  Electroslag, 78, 79, 248
  Field weld, 192